# Wireless Networking
# Visual™
# Quick Tips

### Visual®

## Read Less - Learn More®

## by Rob Tidrow

Wiley Publishing, Inc.

# Wireless Networking
# VISUAL™Quick Tips

Published by
Wiley Publishing, Inc.
111 River Street
Hoboken, NJ 07030-5774

Published simultaneously in Canada

Copyright © 2006 by Wiley Publishing, Inc.,
Indianapolis, Indiana

Library of Congress Control Number: 2006926468

ISBN-10: 0-470-07808-1

ISBN-13: 978-0-470-07808-2

Manufactured in the United States of America

10 9 8 7 6 5 4 3 2

1K/QX/QX/QW/IN

## Trademark Acknowledgments

## Contact Us

For general information on our other products and
services contact our Customer Care Department within
the U.S. at 800-762-2974, outside the U.S. at 317-572-
3993 or fax 317-572-4002.

For technical support please visit www.wiley.com/
techsupport.

WILEY

Wiley Publishing, Inc.

**US Sales**

Contact Wiley
at (800) 762-2974 or
fax (317) 572-4002.

# Prai

*"I have to praise you and your company on the fine products you turn out. I have twelve Visual books in my house. They were instrumental in helping me pass a difficult computer course. Thank you for creating books that are easy to follow. Keep turning out those quality books."*

Gordon Justin (Brielle, NJ)

*"What fantastic teaching books you have produced! Congratulations to you and your staff. You deserve the Nobel prize in Education. Thanks for helping me understand computers."*

Bruno Tonon (Melbourne, Australia)

*"A Picture Is Worth A Thousand Words! If your learning method is by observing or hands-on training, this is the book for you!"*

Lorri Pegan-Durastante (Wickliffe, OH)

*"Over time, I have bought a number of your 'Read Less - Learn More' books. For me, they are THE way to learn anything easily. I learn easiest using your method of teaching."*

José A. Mazón (Cuba, NY)

*"You've got a fan for life!! Thanks so much!!"*

Kevin P. Quinn (Oakland, CA)

*"I have several books from the Visual series and have always found them to be valuable resources."*

Stephen P. Miller (Ballston Spa, NY)

*"I have several of your Visual books and they are the best I have ever used."*

Stanley Clark (Crawfordville, FL)

*"Like a lot of other people, I understand things best when I see them visually. Your books really make learning easy and life more fun."*

John T. Frey (Cadillac, MI)

*"I have quite a few of your Visual books and have been very pleased with all of them. I love the way the lessons are presented!"*

Mary Jane Newman (Yorba Linda, CA)

*"Thank you, thank you, thank you...for making it so easy for me to break into this high-tech world."*

Gay O'Donnell (Calgary, Alberta, Canada)

*"I write to extend my thanks and appreciation for your books. They are clear, easy to follow, and straight to the point. Keep up the good work! I bought several of your books and they are just right! No regrets! I will always buy your books because they are the best."*

Seward Kollie (Dakar, Senegal)

*"I would like to take this time to thank you and your company for producing great and easy-to-learn products. I bought two of your books from a local bookstore, and it was the best investment I've ever made! Thank you for thinking of us ordinary people."*

Jeff Eastman (West Des Moines, IA)

*"Compliments to the chef!! Your books are extraordinary! Or, simply put, extra-ordinary, meaning way above the rest! THANKYOU THANKYOU THANKYOU! I buy them for friends, family, and colleagues."*

Christine J. Manfrin (Castle Rock, CO)

# Credits

**Project Editor**
Maureen Spears

**Acquisitions Editor**
Jody Lefevere

**Product Development Supervisor**
Courtney Allen

**Copy Editor**
Kim Heusel

**Technical Editor**
Justin Kamm

**Editorial Manager**
Robyn Siesky

**Business Manager**
Amy Knies

**Editorial Assistant**
Laura Sinise

**Manufacturing**
Allan Conley
Linda Cook
Paul Gilchrist
Jennifer Guynn

**Indexer**
Infodex Indexing Services, Inc.

**Book Design**
Kathie Rickard

**Production Coordinators**
Adrienne Martinez
Jennifer Theriot

**Layout**
Jennifer Click
Amanda Spagnuolo

**Screen Artist**
Jill A. Proll

**Illustrators**
Ronda David-Burroughs
Cheryl Grubbs

**Cover Design**
Anthony Bunyan

**Proofreader**
Broccoli Information Management

**Quality Control**
John Greenough
Robert Springer

**Vice President and Executive Group Publisher**
Richard Swadley

**Vice President and Publisher**
Barry Pruett

**Composition Director**
Debbie Stailey

# How To Use This Book

*Wireless Networking VISUAL Quick Tips* includes 100 tasks that reveal cool secrets, teach timesaving tricks, and explain great tips guaranteed to make you more productive as you network wirelessly. The easy-to-use layout lets you work through all the tasks from beginning to end or jump in at random.

## Who Is This Book For?

If you want to know the basics about wireless networking, or if you are a visual learner and want to learn shortcuts, tricks, and tips that let you work smarter and faster, this book is for you.

## Conventions Used In This Book

### ❶ Introduction

The introduction is designed to get you up to speed on the topic at hand.

### ❷ Steps

This book uses step-by-step instructions to guide you easily through each task. Numbered callouts on every screen shot show you exactly how to perform each task, step by step.

### ❸ Tips

Practical tips provide insights to save you time and trouble, caution you about hazards to avoid, and reveal how to do things with your wireless network that you never thought possible!

In addition, this book contains pages that define terms and illustrate basic concepts so that you can fully understand everything there is to know about wireless networking.

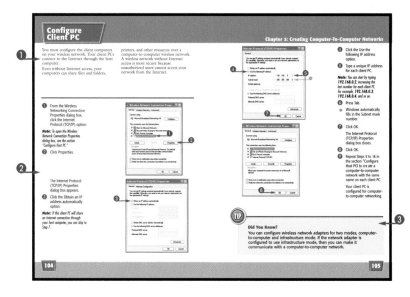

# Table of Contents

 **Introducing Wireless Networking**

 **Setting Up Wireless Network Hardware**

**Installing Wireless Hardware in PCs**

 **Configuring Wireless Networks**

**Creating Computer-to-Computer Networks**

 **Working on Wireless Networks**

 **Administering Wireless Networks**

## Securing Wireless Networks

## Connecting on the Road

## Index                                                                 192

# Introducing Wireless Networking

Wireless networks allow you to connect computers together and access the Internet without using wires or cables. You can use wireless networks in many different locations, businesses of all types and sizes, the military, schools, churches, and municipalities.

Over the past ten or so years, wireless networking technology has improved so much that workers are now mobile, yet they can stay connected to the workplace networks. This ensures that workers can remain in contact with other workers through e-mail, Web sites, file sharing, and shared calendaring even while physically away from the office.

Another interesting facet of wireless networking is a technology called *mesh networking*, which allows a wireless network to "cover" an entire municipality. With a mesh network, all users in the area can access the wireless network as long as they have an account to access the network. This means that a city, such as Atlanta, can offer wireless networking services to its general population.

In this chapter, you learn about the different types of wireless networks and other important issues, such as security concerns.

# Quick Tips

# Discover Wireless Networks

A wireless network enables a group of connected computers and devices to communicate without being physically connected to a network. It eliminates the cables used in wired networks. The most popular wireless networks are called *Wi-Fi*, or Wireless Fidelity, networks.

One of the main problems in the past with wireless networks has been the speed with which users could communicate with each other. Because of new standards and emerging technologies, wireless networks now have similar connectivity speeds as hard-wired networks.

## Radio Signals

Wireless networks use radio signals, similar to those in radio and television broadcasting, to transmit data between devices. Wi-Fi networks operate on the 2.4 GHz or 5 GHz frequency band. These networks can send data at speeds up to 54 Mbps (megabits per second).

## Radio Transceivers

A radio transceiver sends and receives radio signals. Each device in a wireless network has a radio transceiver to send and receive information to and from the network. A transceiver can be located inside or outside a computer.

### Mobility of Network

You can move laptop computers and other wireless-enabled devices while remaining connected to the network. Depending on the technology and other factors, a wireless network has a range of 150 to 350 feet. In addition, as a laptop computer user, you can connect to other wireless networks while traveling.

### Speed

The faster the network speed, the faster that files and other data move from one computer to another computer. Newer wireless network technologies enable faster data transmissions than some wired networks. However, Fast Ethernet and gigabit Ethernet networks can move at least twice the data of the fastest wireless network technologies.

### Cost

Prices for wireless networking equipment are rapidly falling, making it possible to create an inexpensive, fast, and reliable wireless network. While wired networks can be much faster, they often involve intrusive wiring throughout a home or office.

A network is a group of connected computers and devices that allow people to share information and equipment.

A network comprises many different components. Some of these components include a network adapter (also called a network interface card, or NIC), cables, routers, servers, hubs, switches, and network operating systems.

For wireless networks, you can have the same components, except the cables are not necessary. However, most wireless networks do include some wired components for speed or cost concerns. For example, a large company connects its regular desktop computers to a local area network (LAN) using conventional wired technologies. Users who roam around the building or company campus, on the other hand, may rely on laptops or other mobile devices that can access the company-wide network using wireless technologies.

## Communication

Networks enable different computers and devices to exchange information such as files and documents. If a network is connected to the Internet, the network also enables connected devices to access information available on the Internet. A network can consist of many different components or as few as two.

## Infrastructure

The *infrastructure* of a network is the term used to describe the physical bits and pieces across which information travels. Cabling, routers, hubs, and switches are all considered part of the network infrastructure. A small home network can consist of very few components, while large networks can consist of thousands of pieces of equipment and require a full-time team of personnel to maintain them.

## Access Points

*Access points* are the locations on a network that provide access to the network for devices and computers. When used with wireless technology, a network can use a single access point to allow multiple wireless devices to access the network.

## Servers

*Servers* are computers that are dedicated to performing one or a few tasks on a network. Most business networks use dedicated servers for services such as file storage, Internet access, and running applications such as a database program. It is possible to run multiple servers on one physical computer. For example, a single computer on a network may be an e-mail server and a Web server at the same time.

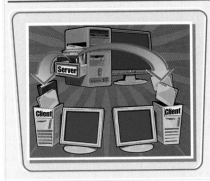

## Clients

*Clients* are computers that require services from the network. Most computers require communication with network servers using applications called client applications. For example, a Web browser is the client application for accessing information from a Web server.

continued

## Peripherals

A *peripheral* refers to a device you can connect either directly to the network or to computers that are connected to a network. Once you have connected a device to a network, anyone on the network can access the device with the appropriate authorization. Some examples of the numerous types of devices you can connect to a network are printers, scanners, storage devices, and cameras.

Star Topology

## Topology

The *topology* of a network determines the physical layout of the network. The most common topology is the star topology that has a device such as a hub or router as the center of the *star*, connected to different devices. Other lesser-used topologies are the *bus* and the *tree* topologies.

## Backbone

The *backbone* of a network is the term used to describe the main cable within a network where most network traffic traverses. A backbone is typically connected to many other devices such as routers and switches, rather than to single computers. The backbone of a network must be able to handle large amounts of information or bandwidth. Large networks typically use fiber-optic cables and very fast network devices.

BACK BONE

## Network Protocols

Computers and other devices connected to a network can communicate with each other because they all agree to use the same method of exchanging information called a *protocol*. A network can use many different protocols simultaneously. For example, a Web browser communicates with a Web site using a protocol that specifies how Web information is exchanged, while two network cards exchange messages over a cable using a different protocol that dictates how information is transmitted via electrical signals on a cable.

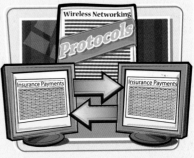

## Cables

Apart from very simple wireless networks, all networks contain cabling, which exchanges information between computers and devices such as switches. The most common type of cable is *twisted-pair* cable, which contains four pairs of wires (eight wires total) that are entwined with each other. The end of the cable terminates with a connector in the shape of a large telephone connector.

## Size

The size of a network determines how the network is referenced. A network within a single building is called a *local area network*, or *LAN*. Networks that connect across a larger area or even across a country are called *wide area networks*, or *WANs*. LANs that contain wireless technologies are referred to as *Wireless LANs*, or *WLANs*.

# Discover the Benefits of Wireless Networking

Because of the wide range of benefits wireless networks provide, both business and home users are using networks more and more.

The main benefit of a wireless network is how easily users can become mobile and still access network resources at the same time. A corporate manager, for example, can carry her laptop computer to a scheduling meeting and access e-mails, shared calendars, and other resources using the wireless network, eliminating network cable connections.

Another way workers can become mobile is using handheld devices, such as a personal digital assistant (PDA). PDAs are small devices that the user can carry in a pocket, purse, or hip holder, and that include software for accessing e-mail, Web services, shared files, and more. A common use of PDAs is accessing e-mail accounts while away from the office. A worker can connect to the wireless network, download unread e-mails, create new e-mails, read and respond to messages, and manage e-mail content.

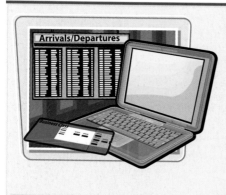

## Remain Mobile

One of the major benefits of wireless networking is being able to remain mobile while using a computer, but still have access to all the services and resources made available from a network, such as the Internet. You can even use a laptop computer while you move in a car or plane, as long as the computer is in range of the wireless network.

## Fast Setup

Once you set up the infrastructure for a wireless network, you can add more computers and devices to the network quickly. Once you add a wireless network adapter to a computer and configure the computer to use wireless networking, the computer can connect to the network immediately.

### Cost

As opposed to networks that use cables, wireless networks can be much cheaper to set up. Apart from the costs associated with equipment like hubs and repeaters, the installation of cable through an existing building may be very difficult and expensive. Wireless networks also allow networks to travel across objects, such as roads, that require a lot of work and money to cross with cables.

### Easy to Move

In traditional networks that require the use of cable, you cannot easily move a computer to a new location, because you must attach the computer to a nearby outlet using a cable. It is sometimes difficult to move a network computer to a new location within the same room. A wireless network allows you to move any computer anywhere, as long as the computer is in range of the wireless network.

### Expandability

Adding a new computer or device to a wireless network is as simple as turning the computer on. Most wireless devices, such as access points, can support many different devices, and as long as you do not exceed the maximum number of devices, the access point quickly accepts new connections. If needed, you can add multiple access points to a wireless network to facilitate large numbers of computers.

# Understanding the Disadvantages of Wireless Networking

While wireless networks have a wide range of benefits, there are also unique disadvantages, some of which include speed, battery life of mobile devices, interference, security issues, and cost. Another problem with wireless networks is interruptions in services. Companies and cities that rely on wireless networks have to test for and fix problems in areas where wireless services are interrupted due to buildings blocking transmissions or stronger radio signals interfering with transmissions.

A major concern for companies that deal in personal information and data is wireless networking security. Because wireless networks use radio frequencies, it is difficult to control access to these frequencies. In fact, anyone in the area with a wireless device can access the frequencies. This means the company must use access privileges to control access to a network via a wireless device. Only those people and devices with authorized usernames, device IDs, and passwords are allowed on the network.

## Power Consumption

Each wireless device in a computer, such as a laptop or a handheld computer, has a radio transmitter and receiver. Radio devices require a relatively large amount of power to operate effectively. Using wireless adapters on portable devices can greatly reduce the length of time that the devices can operate using battery power.

## Interference

Wireless networks use radio signals to transmit information. Unfortunately, there are many types of devices that use radio waves to operate. These other devices can interfere with the signals that the wireless network uses. Tracking down and eliminating interference sources can be difficult.

### Network Security

By their very nature, wireless networks are more susceptible to unauthorized access. A network may be accessible from a location not under the control of a network administrator, such as a parking lot next to the building housing the wireless network. While cable networks have the same concerns, they are not as easy to access as wireless networks.

### Inconsistent Connections

With cable networks, computers are ensured a direct, stable connection to the network. However, moving a computer to another location or items blocking the path of transmission can interrupt wireless network connections. While many applications, such as Web browser applications, are adversely affected by temporary connection loss, other applications, such as database-based applications, may result in information loss.

### Lack of Management

With a wired network, network administrators can exercise very tight control of the physical components of the network. For example, network administrators can ensure that all cables are the correct distance from devices that may cause disruptions, such as light systems or photocopiers. With wireless networks, short of physically inspecting each wireless device, there is no way that administrators can determine or control the exact physical layout of the network.

# Discover the Types of Wireless Technology

Wireless technology enables computers and devices to communicate with each other without the use of wires. There are many different types of wireless technologies, each one with its own set of strengths and weaknesses.

Companies or municipalities may settle on one type of technology, such as infrared, for their network. Or they may employ two or more types of wireless technologies depending on their needs. For example, a company that has a campus setting with several buildings may use microwave and infrared. The microwave technology can transfer data from building to building, while you can use the infrared devices inside the building.

## Wi-Fi

*Wireless Fidelity,* or Wi-Fi, is becoming the preferred technology for creating wireless networks both at home and at work. Wi-Fi allows computers and devices, such as printers and hubs, to communicate without using cables. Most new wireless networking devices in use are Wi-Fi devices. Wi-Fi is also used to facilitate Internet access in public places, such as airports.

## Bluetooth

*Bluetooth* is the name of the wireless technology that is used primarily to allow individual devices to communicate with each other over short distances. For example, handheld computers can transmit a phone number from an address book to a mobile phone, which then dials the number. While it is possible to use Bluetooth to network computers together, this is not generally done.

## Infrared

*Wireless infrared technology* allows two devices to communicate using infrared light and you most commonly find them in remote controls. Infrared devices need to maintain a constant line of sight between the devices and are more reliable over short distances. The most common use of infrared technology is allowing handheld computers to exchange data with each other and with laptop computers. Most handheld devices and laptops have a built-in infrared port.

## Cellular

*Cellular wireless technology* is most commonly associated with mobile telephones. Each telephone communicates with a nearby transmitter, which changes as the phone moves around a location. Laptop computers routinely use cellular phones as modems to provide dial-up access from remote locations.

## Microwave

*Microwave technology* enables two devices to communicate using microwave dishes that are aligned with each other. You can use microwave systems to connect the networks of two buildings that are separated by obstructions such as wide roads. Microwave systems are very expensive but can transfer large amounts of information.

You can increase the efficiency of many specific applications using wireless networking.

By far, the main application for wireless networking is person-to-person communication. All types of users use e-mail and messaging applications, from the president of a Fortune 500 company sending quarterly company results, to a youth baseball coach sending messages to players about upcoming practices.

Another application more and more in common use in wireless networking is scheduling software. With scheduling software, groups of people, departments, and families can share calendars to help manage meeting times and appointments.

## E-mail

E-mail is by far the most popular networking application, and wireless networking now makes it possible to access e-mail constantly. You can use laptop computers, and, increasingly, handheld computers, to access e-mail wirelessly at work and elsewhere with more frequency at public locations, such as airports and cafes.

## Messaging

Most operating systems provide a messaging application that allows you to communicate instantly using text. Even inexpensive handhelds now have the capability to provide messaging services and, when coupled with wireless networking, allow you to stay in constant communication with your colleagues and friends wherever you are.

## Scheduling

The ability for people to schedule activities and notify others of their activities greatly increases the efficiency with which they can work together. Allowing workers to update their schedules and exchange that information immediately using wireless technologies only further increases the efficiency of the scheduling system.

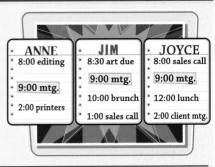

## Data Collection

Making computers mobile increases the speed at which you can update data in applications such as inventory control systems. For example, a person in the warehouse can immediately update information about the number of products on a shelf instead of waiting until one has access to a computer.

## Web Browsing

For most home users, the primary use of wireless networks is the ability to use a mobile computer, such as a laptop, to connect to the Internet, regardless of where they are in the house. Most wireless networks used at home are easily connected to the Internet and can provide access not only in the house, but also in the area outside of the house, such as a deck or patio.

To connect to a wireless network, you must equip your computer with minimal hardware and software. Doing so makes your network faster and more dependable.

Another reason to review system requirements is to ensure your computer, handheld device, or other component meets the wireless networking standards that your company, school, or home uses.

For example, to communicate with other computers on a wireless network, your computer must use the same wireless standard as another computer on the network. This is analogous to you speaking English and your friend speaking Spanish; one or the other must be able to understand and speak the other person's language or your communication shuts down.

## Operating System

You need a PC running Microsoft Windows XP to follow what is in this book, either the Home or the Professional version. Windows XP is an operating system that controls your computer. Windows XP has built-in wireless networking support that makes it easy to create and maintain wireless networks. However, you do not need Windows XP to go wireless.

## Hardware

You must equip your desktop PC or a laptop computer with a Pentium 4 1.2 GHz processor, 256 megabytes of memory, and a 40GB hard drive.

## PC Ports

You need to install some basic networking equipment on your computer. To start, your computer needs some empty slots where the equipment is connected. Your desktop PC should have an available PCI slot. Your laptop computer should have a PC card slot. If neither of these is available, it is also possible to use the USB port with some equipment. For more information about attaching networking equipment to your computer, see Chapter 3.

One advantage of using USB ports for networking equipment is the ease at which you can connect the equipment. With USB you simply plug in the equipment, such as a network interface adapter. You do not have open up the computer case to make the connection.

## Internet Access

If you want to connect to the Internet with your wireless network, you need Internet access. To get the full benefit of a wireless network, you need high-speed, or broadband, access. This is available through a local telephone company's DSL service or a cable TV system's cable modem service. You can access the Internet directly or through another computer if that computer is using the Internet Connection Sharing feature available with Microsoft Windows XP.

# Consider Your Networking Requirements

Before deciding the type of wireless network to build, consider how many computers will connect to the network and what operations you want the network to provide.

One way to know what your wireless network needs is to take inventory of the types of services, applications, and connections you use now, and then adapt those things that can go wireless into your wireless network design. For example, if you currently use e-mail, scheduling, data warehousing, and spreadsheet applications, consider making all those available over the wireless network. You can store your spreadsheets, for example, on a central server that you can access by wired and wireless computers.

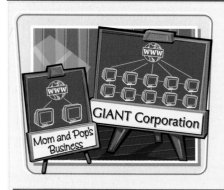

## Set Up Home or Office

The simplest wireless network is one home computer connected to a broadband Internet connection. In a larger family, there may be several desktop and laptop PCs connecting to the Internet and to each other through a wireless network. In an office environment, you may want to choose one of the newer, faster networking technologies that use the 802.11g standard.

## Number of Computers

The number of desktop PCs, laptop computers, and other network-enabled devices determines how much wireless networking equipment like routers and network interface cards — NICs — you need to purchase, install, and configure. Your wireless network may consist of one to dozens of computers connecting to the Internet and to each other.

## Mobile Access

If you want to access wireless networks while traveling, you need a laptop PC or personal digital assistant, or PDA. These devices let you connect to available Wi-Fi hot spots in many places you visit. You also can use the devices to connect to your wireless network when you are at home or in your office. Laptops come in several different varieties, including lightweight, durable, high-performance, large screen, and so on. The one you pick should match the type of work you plan to do. For example, if you travel a great deal and want a lightweight model, look for one that weighs 4-6 pounds. However, if you need a high-performance laptop that can handle graphics-intensive software, you may need to opt for a heavier laptop that weighs over 10 pounds.

## Mixing Wi-Fi Standards

If you connect computers and equipment that use multiple standards, you need *dual-band* capability. This allows you to mix and match wireless technologies. For example, if one network uses equipment with the 802.11a standard and another operates under the 802.11b standard, you can connect them together using a dual-band router.

The most popular wireless networking technologies today are based on the 802.11 standard, which governs how devices on the network communicate with each other. The Institute of Electrical and Electronic Engineers (IEEE) developed the standard, popularly known as Wi-Fi, of which there are several variations.

When you set up purchasing requirements for your home or office wireless network infrastructure, take the time to understand the wireless standards that each component supports. If you use Wi-Fi, for example, all your components that connect to that Wi-Fi device must be Wi-Fi compatible.

### Wi-Fi

Wi-Fi stands for *Wireless Fidelity*. It now generally refers to all the 802.11 wireless networking standards, which specify how devices communicate using wireless networks, although it originally identified networks that used the 802.11b standard. The Wi-Fi Alliance, a nonprofit industry association, works to ensure interoperability among the various 802.11 wireless technology standards.

### 802.11a

The 802.11a standard is the least popular of the Wi-Fi technologies. While it is beneficial for some office networks with high-bandwidth needs and closely located computers, it has a short range. Dual-range equipment allows 802.11a equipment to network with the more popular 802.11b standard and the newer 802.11g standard. Otherwise, the 802.11a standard is incompatible with the 802.11b and 802.11g standards.

## 802.11b

The 802.11b standard is the most popular of the Wi-Fi technologies. It transmits data at a slower speed than both the 802.11a and 802.11g standards. Unlike 802.11a networks, 802.11b radio waves can penetrate most walls but are susceptible to interference from cordless phones, baby monitors, and microwave ovens.

## 802.11g

The 802.11g is the newest Wi-Fi standard with the same range as the 802.11b standard but with the ability to transmit data at a much faster rate. It can communicate with 802.11b networks, but requires dual-band equipment to interact with networks based on the 802.11a standard. It suffers from interference problems similar to networks using the 802.11b standard, such as cordless phones, baby monitors, and microwave ovens.

## 802.11i

The 802.11i is an emerging standard that will increase the security of Wi-Fi networks. When it is available, you may be able to upgrade some older equipment to this newer standard.

# Discover Network Configurations

You must choose which type of wireless network configuration you need. The two general types are infrastructure and computer-to-computer, or ad hoc.

The *configuration* is the way in which the network is laid out to allow all computers on the network to communicate with each other and the servers. How you configure your network depends on a few factors, including size of network, costs, resources you want to access, and the number of users who will be accessing your network.

Some wireless networks need a centralized access point that several computers will use to "jump" to a larger network (such as the Internet). Other wireless networks may be small enough where a single computer can act as a shared device that can then allow the other computers to access other networks.

## Infrastructure

An *infrastructure* network is the most widely used wireless network configuration. It uses a wireless router, also called an access point or gateway, to connect to the Internet through a broadband modem. The wireless router then communicates with other wireless-enabled devices on the network. Infrastructure networks can bridge wireless networks with existing wired, or Ethernet, networks.

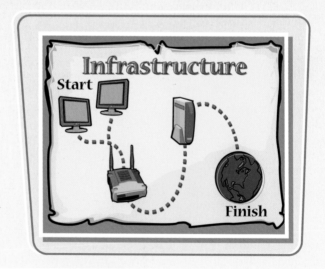

## Computer to Computer

A computer-to-computer, or *ad hoc*, network allows computers to communicate with each other without the use of an access point, such as a router. This basic network configuration permits you to exchange files among computers. In addition, one computer in a network can directly connect to the Internet and permit other computers to share the connection.

Computer to computer networks are the easiest types of wireless networks to create and manage. As long as one computer is set up as a host computer, other computers can connect to the host as client computers. From there, all the computers on the network can communicate with each other, enabling them to share files, share printers, and communicate with e-mail.

# Discover Networking Speeds

The speed of your network depends on many factors, such as the technology you use and the layout of your home or office. You may choose to use a technology that gives you a faster speed but less range.

Unfortunately, there are no wireless technologies available that allow extremely high speeds over great distances. Perhaps in the future there will be technologies that allow users to connect to their offices halfway around the world using laptops and wireless network cards. At the present time, you can be several hundred feet away and get high-speed connections. However, as your distance increases, your speeds deteriorate and eventually fade out.

## Reality Check

You will find that theoretical network speeds and real network speeds are not the same. Your wireless network's actual speed — that is, the rate at which it transfers data — depends on the distance between computers, which standard you choose, and the manufacturer of your equipment.

## 802.11a Speed

The 802.11a standard has a potential speed of 54 Mbps.

### 802.11b Speed

The 802.11b standard, the most popular wireless networking mode, has a potential speed of 11 Mbps.

### 802.11g Speed

The 802.11g standard, while operating in the same frequency band as the 802.11b standard, has a potential speed of 54 Mbps.

### Ethernet Speeds

In contrast to wireless networking speeds, wired Ethernet networks are still the speed winners. Ethernets operate at only 10 Mbps, less than Wi-Fi speed. However, Fast Ethernets operate at 100 Mbps, and the newer gigabit Ethernets are even faster. While you gain in speed with the faster wired networks, you lose the freedom of mobility that Wi-Fi technology provides.

The technology you use and your environment determine how far your wireless network can reach. Walls and metal structures reduce the range.

Planning for the coverage range of your wireless network can be an inexact science. You can measure the distance between buildings, compare your location with others in your area, and research all the standards out there. However, you will not know exactly how far your wireless network reaches until you install it and start working on it. In some cases, you may end up relocating wireless switches and servers to eliminate or reduce obstacles in their way.

### 802.11a Range

Operating in the 5 GHz frequency range, the 802.11a standard is best suited for dense networks with high bandwidth needs. In a typical office environment, 802.11a networks have a possible range of up to 255 feet. The typical range is 25 to 75 feet indoors. Coverage is limited to one room.

### 802.11b Range

Wireless networks operating on the 802.11b standard have a greater range than those using 802.11a. As the most popular standard, it often is used for public access locations, or hot spots. Its signals, operating in the 2.4 GHz frequency band, penetrate most walls and have a possible range of up to 300 feet. The typical range is up to 100 to 150 feet indoors.

## 802.11g Range

802.11g, the newest standard in the Wi-Fi networking family, has the same range as the 802.11b standard: up to 300 feet, but typically 100 to 150 feet indoors. However, it is much faster than networks using the 802.11b standard. Operating in the 2.4 GHz frequency range, 802.11g network signals can penetrate most walls.

This allows homes and small businesses to add 802.11 g devices in central locations and provide wireless access to most of their users. In addition, because of the high connection speeds of 802.11 g devices, multiple computers can connect to the same 802.11 g device (such as a router) without noticeable performance issues.

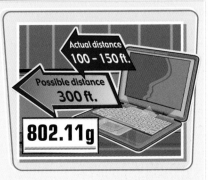

Actual distance
100 – 150 ft.

Possible distance
300 ft.

802.11g

Baby

## Other Range Factors

Some cordless phones, baby monitors, and microwave ovens can interfere with Wi-Fi networks using the 802.11b and 802.11g standards, decreasing their range. Keep this in mind when you position your wireless networks around children's rooms, kitchens, and break rooms. You may also find that cellular phone towers interfere with your wireless devices. If you are near one of these towers and interference is an issue, consider using a combination of conventional wired devices and wireless devices for your network.

# Setting Up Wireless Network Hardware

Before you create a wireless network, you should know about the different types of networking hardware. This chapter helps you install and configure routers and other wireless networking hardware.

In many respects, wireless networking hardware is the same as conventional wired networking hardware. The main difference is that the wireless hardware uses devices that can receive and transmit radio waves. These waves deliver packets of digital data.

The nice part about many of the devices that wireless networks use is that all that you usually need is a nearby electrical outlet. You do not have to worry about pulling cable through ceilings, walls, or crawl spaces to get your network up and running.

*Quick Tips*

# Understanding Wireless Hardware Installation Issues

There are a number of concerns that you need to address before installing and setting up wireless hardware.

One point to remember is that regardless of the equipment, always refer to the equipment documentation before installation and setup.

Also, when you install wireless network hardware, try to anticipate where users will most likely congregate to access the network. For example, if you plan to include wireless access points in conference rooms, place the devices in the most centralized location in the room. Do not hide them in a corner or place them in the middle of the conference table. Consider attaching them to the ceiling of the room so they are up and out of the way, but so that users can access them.

## Location

When installing wireless hardware, pay special attention to the location of the equipment. For example, placing a wireless access point in the middle of a basement may give better coverage throughout the house than if you place the access point in the corner of the basement. If you expect to use the upper floors in the house frequently, you may consider placing the access point on the first floor instead of the basement.

## Cabling

Some wireless equipment, such as routers, can have standard Ethernet network ports on them to enable a computer to connect directly to the wireless device by using a network cable. If you plan to install a wireless device next to a computer that will use the device, look for wireless equipment with a network port so you can use standard, more inexpensive, cable rather than using wireless technology to span a short distance.

## Ease of Access

You can position most network equipment in many ways, such as fastening it to a wall or stacking it on top of other network equipment. Unlike other nonwireless network devices, you may have to move wireless network equipment and even adjust the antenna from time to time as the wireless network changes. When positioning your network hardware, always give yourself plenty of room to work with the equipment.

## Internet Access

Some applications, such as those that update the firmware on a hardware device, require that you have access to the Internet in order to obtain the necessary files to update the equipment. Before setting up your wireless device, check the hardware manufacturer's Web site to ensure that you have the latest version of any software to configure your hardware.

## Available Computer

Almost all wireless equipment requires the use of a computer to set up the initial hardware. Some devices require a connection to a computer with a network card and a network cable. The configuration software can then run on the computer, and you can use the software to adjust the settings on the wireless device. Once you configure the wireless device, you can detach the computer from the device.

## Settings

There are many different network settings that you may need to know to configure your wireless equipment; for example, computer names and IP addresses. Before setting up your wireless equipment, you should have available all the necessary information, including the Internet settings of your ISP if you want to connect the network to the Internet.

At a minimum, your wireless infrastructure network requires a broadband modem and wireless residential gateway. Choosing the correct wireless hardware allows you to build a wireless network that fits your needs now and in the future.

Consider investing in backup or duplicate hardware in some instances. For example, some companies purchase additional wireless network adapters for their cache of laptops. This allows them to quickly replace any that have been stolen, lost, or damaged. It also provides duplicate hardware for new laptops they may purchase in the future. The IT department will already be familiar with the brand and model of the adapter, so installation and configuration for the newly acquired laptops goes more smoothly.

## Broadband Modem

A *broadband modem* connects your wireless network to a high-speed Internet connection. Your broadband provider, which is usually your telephone or cable TV company, normally installs and configures this hardware for you. Some residential gateways include broadband modem functionality.

## Residential Gateway

A wireless *residential gateway* is a hardware device that handles data traffic between your broadband modem and the computers on your network. A wireless residential gateway, which is one type of wireless access point, incorporates a *wireless router* that transmits and receives radio signals to and from wireless devices on your network. Wireless infrastructure networks require either a residential gateway or a wireless access point.

### Access Point

You can use a wireless *access point*, or AP, to extend the range of your wireless network and fill in radio-signal coverage gaps. You also can use a wireless access point to connect a wireless device to your wired network without purchasing a residential gateway or router.

### Network Adapter

Each computer or device on your wireless network requires a network adapter. Network adapters are also known as network interface cards — or NICs. For more information on network adapters, see Chapter 3.

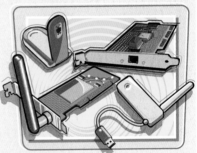

### Network Bridge

You typically use a *network bridge* to connect two networks together. You can also use a bridge to connect a device with a network port, such as a network printer, to an existing wireless network. You do not typically use network bridges in home networks.

continued

## Laptop Adapters

Most laptop computers have an expansion slot that you can use to add features such as a modem or network adapter to the laptop. The PC card slot on a laptop, sometimes referred to as the PCMCIA slot, can accept PC card wireless network adapters. Unfortunately, you must remove any existing PC card in order to install the wireless network card. Many newer laptops have a wireless interface built into them, leaving your PC card slots open and available for use for other accessories.

## Handheld Adapters

Many handheld devices have a built-in slot called a *compact flash interface*. You can insert a network interface card in the slot in the same manner as inserting a card into a laptop. Despite the many uses of the compact flash slot, you must remove all cards, including memory cards, from the slot to use wireless networking. Although similar in size and shape to network cards in laptop computers, wireless network adapters in handheld computers are not compatible with laptop computers.

## External Adapters

Wireless adapters are also available that you can connect to the USB port of a laptop or desktop computer. A USB wireless adapter makes it easy to move a network interface to multiple computers as needed. Most computers, including laptop computers, have a USB port that you can use to connect to a wireless adapter.

### Built-In Wireless Networking

As wireless networks become more popular and standards become more accepted, more computer devices have wireless networking built in. Currently, most major manufacturers of handheld and laptop computers offer models with built-in wireless networking.

### Game Adapters

Game consoles are dedicated gaming systems used in the home solely for playing games. You usually attach them to a television. Many game consoles have network connections that enable users to play games via the Internet or against users on another game console on the network. Some manufacturers are now creating wireless network adapters that are optimized for use with game consoles.

### Personal Video Recorders

*Personal Video Recorders* (PVRs), sometimes called *Digital Video Recorders*, are the new generation of devices that record TV for later viewing. These devices are the next generation of VCRs. Some of these systems, such as certain models of TiVo, now offer wireless network capability to expand the abilities of the system.

# Configure Broadband Modems

Most wireless networks are connected to a high-speed Internet connection using broadband modems, which are not really like the modem you use to connect your computer to a telephone line. Instead, cable modems convert (modulate and demodulate) cable television signals so your computer can access the Internet over cable.

You can purchase broadband modems, but most cable and DSL providers offer them as lease options when they set up your Internet connection. The advantage of leasing is you get updates to the equipment as they become available. The disadvantage is you may end up paying more for the cost of the cable modem over the long run.

## Cable Modems

*Cable modems* provide a high-speed Internet connection using the same cable line that brings the television signal into a building. Your local cable television provider supplies cable modems. Cable modems typically operate at least 10 times faster than a dial-up modem. Cable modems also provide *always-on* Internet access, meaning that you do not have to wait before using the Internet.

## DSL Modems

*Digital Subscriber Line* (DSL) is a high-speed connection that your telephone company provides using your existing telephone line. The telephone company makes minor modifications to your existing phone line that allow you to send digital signals over the line along with your telephone signals without affecting the quality of your telephone calls. DSL, like cable modems, is usable while you are on the telephone and is *always-on* Internet access.

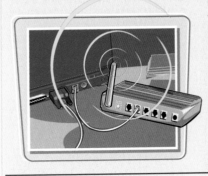

## Connections

Traditional broadband modems connect to wireless equipment using a standard Ethernet network cable. Some broadband modems will only connect to a computer using a USB port. You cannot connect these modems directly to a wireless network. Newer broadband modems have built-in wireless and wired capabilities, which provide a single device that serves as your broadband modem, an Ethernet switch, and your wireless access point.

## Number of Users

Some providers of high-speed connections that use broadband modems, such as cable or phone companies, may restrict you in the number of users that can use the connection. Always check with your connection provider to ensure that you can use the high-speed connection with a network, whether it is a wireless network or not.

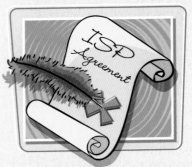

## Authentication

Broadband modems can use a variety of methods to authenticate a user before one can use the modem. Some broadband modems require connection to a computer with a specific name, while other modems use a name and password login. Wireless equipment that is attached to broadband modems can use these authentication methods.

# Configure Network Bridges

Network bridges are simple devices that expand the coverage area of a wireless network, allowing two separate networks — even if the networks are different types and in different locations — to work as one.

With a bridge, data flows from one network to another regardless of whether one network is sending TCP/IP data while the other is sending NetBEUI data. This is helpful when tying together two different networks.

You must configure the receiving computer to read the data it receives by setting up additional protocols. In addition, you must configure network interface cards to promiscuous mode; that is, the NIC listens to all traffic.

## Ethernet Devices

A network bridge enables you to connect any Ethernet device to an existing wireless network quickly and easily. Some devices only have a built-in Ethernet port with no other means of connecting to a network, wireless or otherwise. A wireless bridge enables a device that would otherwise require an Ethernet cable to communicate with a wireless network.

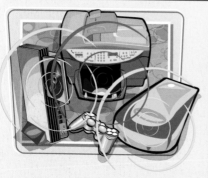

## Devices

Computers and laptops that are commonly connected to a wireless network can easily accept network interface devices such as wireless network adapters or PC card network adapters. You use wireless bridges to connect devices that have only a standard Ethernet port to a wireless network. Examples of devices that only have an Ethernet port are network printers, network storage devices, and even some computer gaming consoles.

## Speed

As with all wireless devices, the speed at which information is transmitted depends on the standards that the device supports. Currently, two popular standards support transmission rates of 11 megabits per second (Mbps) and 54 Mbps. Specialized wireless devices on the market today intended for use by large networks can operate as fast as 480 Mbps. Newer standards increase the speed to 3 gigabits per second (Gbps). The Ethernet port on the

bridge allows a connection to any Ethernet device that uses a connection speed of up to 100 Mbps. Despite the speed at which the bridge communicates with the device, the device can only communicate with the wireless network at the maximum speed allowed by the bridge as it communicates with another wireless device on the network, such as a router.

## Point to Point

It is possible to use two Ethernet wireless bridges to connect two devices or networks together without the use of other wireless devices, such as wireless routers. For example, you can use a network bridge connected to a network in an office to communicate with another network bridge connected to a network located in a warehouse on the other side of a road. This allows both networks to communicate with each other.

## Web-Based Configuration

As with many wireless network devices, a network bridge's configuration is Web based. Regardless of the type of device connection to the bridge, you must connect the bridge to a computer, thereby enabling access to the configuration by using any Web browser.

# Configure Residential Gateways

With a residential gateway, or *router*, you can quickly and easily connect multiple computers to the Internet using a high-speed Internet connection, such as a DSL or cable modem.

When you connect one computer to a high-speed Internet connection, you do not need a router. However, when increasing the number of computers from one to many, you must install a router to manage all the network information between your computers and the broadband modem.

For wireless networks, you must use a wireless router, which usually has at least one built-in Ethernet port for you to plug in a computer to let you set up and administer the wireless router. Once the router is set up, you can switch the computer to a wireless connection.

## IP Address Allocation

All computers that access the Internet, whether connected wirelessly or not, must use a unique identifying number called an *IP address*. A residential router can automatically manage and assign IP numbers to computers connected to the same network to which the router belongs. The management of these IP numbers is accomplished using the *Dynamic Host Configuration Protocol*, more commonly called DHCP.

## Configuration

The configuration of wireless routers is accomplished using any Web browser to access the configuration information of the router.

## Security

Most residential routers now have built-in security features, such as data encryption, and additional software that increase the security of your network. Firewall applications that prevent unauthorized access to the router as well as information monitoring tools are now commonly available with many wireless routers. Some residential gateways even come with parental control software to help restrict access to objectionable content on the Internet.

## Ports

Most residential routers also have a built-in switch, or hub, that allows you to connect computers and other network devices, such as a network printer, directly to the router using standard Ethernet cables. Most routers have four built-in Ethernet ports. Regardless of the speed at which the router operates wirelessly, the built-in Ethernet ports can communicate with each other at speeds of up to 100 Mbps or even 1 Gbps.

## Compatibility

All wireless residential routers can communicate with other devices that use the same communications protocol as the routers. Most new routers support different transmission speeds, enabling you to use a wide range of wireless devices from many different manufacturers running at different transmission speeds.

# Understanding Access Points

*Access points* provide network access for other wireless network devices. You can use access points to allow computers to share Internet connections, extend the reach of an existing wireless network, and provide connections to specific parts of a network, such as a printer bay or training room.

You can use the same type of wireless router you use for setting up a residential gateway as a wireless access point. In fact, many companies purchase the same router model to install as a gateway and access point.

## Provide Access

The primary purpose of an access point is to provide an entry point to the network for other wireless devices. Any computer with a wireless adapter and within range of the access point can connect to the network.

## Extend Coverage

You can use an access point to extend the coverage of a wireless network. For example, if a router is too far away from an area at the rear of a building, you can place an access point in the remote area to bring network access to any wireless devices there.

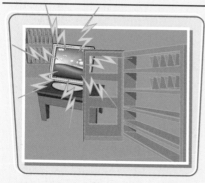

### Limits

Depending on the number of users and the amount of network activity, some access points may have a limit on the number of users that can simultaneously use the network using the access point. Consult with the hardware manufacturer to determine the capabilities of the access point you set up.

### Ports

Unlike routers, some access points typically only have a single network port through which they connect to the network. You can connect the access point to an existing cable network via a hub or switch, or you can connect it to another wireless device, such as a wireless router.

### Compatibility

As with all wireless network devices, make sure the access point can interoperate with the other wireless network devices on the network. Generally, devices from different manufacturers operate together as long as they use the same wireless standard.

# Set Up a Wireless Gateway

You can connect computers to your wireless network after you set up your wireless gateway, wireless residential gateway, or wireless router, which links the internal wireless network with the Internet. You configure your gateway with a Web browser or setup program that your router supplies.

During set up, you may be asked for the IP address of the router. In some cases, your Internet service provider (ISP) gives this to you, but mostly your ISP assigns you an IP address. Depending on your ISP contract, you may need to periodically obtain a different IP address from your provider. Some providers have a 24-hour rule in which you must obtain a different IP address. Others may be longer. The acquisition and setup of these dynamic IP addresses are usually automatic.

---

 In your Web browser, type **http://192.168.2.1** and press Enter.

*Note: The default address of your gateway may be different; your gateway manual provides this information.*

The Base Station Management Tool window appears.

 Type the administration password in the Password box.

 Click Log On.

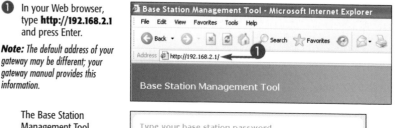

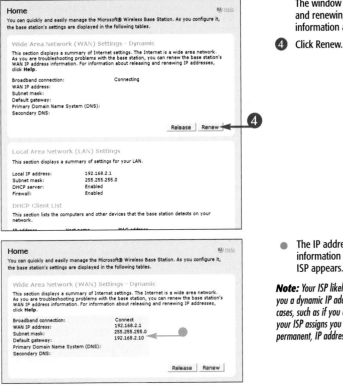

The window for viewing and renewing network information appears.

④ Click Renew.

● The IP address information from your ISP appears.

***Note:*** *Your ISP likely has assigned you a dynamic IP address. In some cases, such as if you are a business, your ISP assigns you a static, or permanent, IP address.*

**TIP**

**Did You Know?**
You should change the security settings of your wireless gateway after you set it up. However, do not enable encryption features on your hardware until your network is working. You can return and enable security features on your routers and computers later. To learn more about wireless network security, see Chapter 8.

You can set up additional wireless *access points*, or APs, to extend the range of your wireless network. You connect an AP to your wired network with an Ethernet cable and then configure the hardware with a Web browser.

To set up your access point, you must run the access point configuration utility.

In most cases, this is a Web page that connects to your access point hardware (router). You can renew IP addresses, change time settings, upgrade the hardware's firmware, change passwords, and configure security settings.

## CHANGE THE PASSWORD

① To change the administration password, in your Web browser, type the address of your access point and press Enter.

② Click Management.

③ Click Change Password.

The access point home page appears.

The Change Password window appears.

④ Type your current password.

⑤ Type a new password.

⑥ Retype the password to confirm it.

⑦ Click Apply.

Your password is changed.

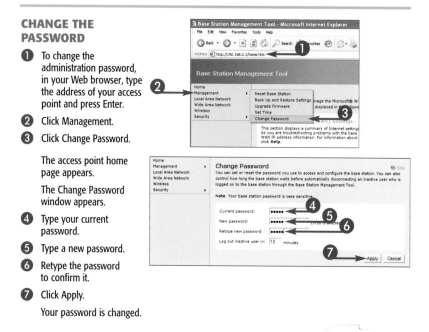

## SET THE TIME ZONE

**8** Click Management.

**9** Click Set Time.

The Set Time window appears.

**10** Click here and select a time zone.

**11** Click Update Time Settings.

The time zone for your access point updates.

### Did You Know?

Manufacturers sell wireless access points that work with two or three different Wi-Fi modes. A dual-mode AP may work with the 802.11a and 802.11b technologies, while a tri-band AP may work with these modes plus the emerging 802.11g standard.

# Download Firmware

You can download updates to the software, or *firmware*, that operates your router or other wireless access point. Firmware updates can improve the functionality of your hardware.

Some firmware updates provide bug fixes to common problems associated with your hardware. For example, one update may fix a bug that prevents more than four

users from connecting to a router simultaneously. Another update may simply change a button on the software.

Not all hardware requires or has firmware updates. However, as new operating systems and wireless standards emerge, manufacturers usually release firmware updates to keep their older hardware up to date.

## DOWNLOAD FIRMWARE

① In your Web browser, type the address of your access point and press Enter.

The access point home page appears.

② Click Management.

③ Click Upgrade Firmware.

The Upgrade Firmware window appears.

④ Click the Microsoft Broadband Networking Web site link.

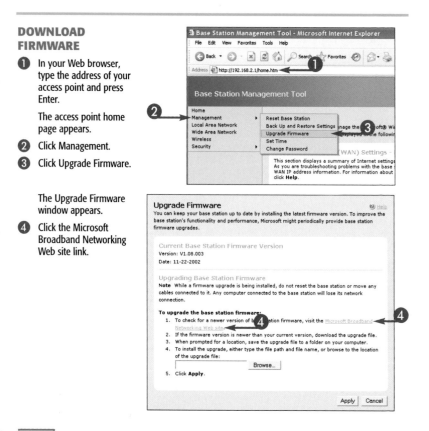

**Broadband Networking**

- Microsoft Help and Support
- Microsoft's support LifeCycle for Broadband Networking products
- Frequently asked questions regarding Microsoft's Support LifeCycle
- Search the Microsoft Download Center for any available downloads
- Update the firmware and software for your Microsoft Broadband networking devices

**Manual update**

1. Download the MSBNDownload.exe firmware update. To do this, the following Microsoft Web site:

   http://go.microsoft.com/fwlink/?linkid=6790

2. Save the file to a location such as your desktop or the My Documents folder.
3. If you are updating a wireless base station, use an Ethernet cable to connect your computer to your base station.
4. To run the update, double-click the **MSBNDownload** file.
5. Follow the instructions that appear on the screen to update the firmware of your device.

While the firmware is being saved to your base station, the power light on the base station blinks and then turns orange. When completed. If the update fails, the power light continues to blink slowly until you successfully update the firmware. In this again, or you can reset the base station.

+ Back to the top

The Microsoft Broadband Networking page appears.

**Note:** *The link you click and visit depends on the manufacturer of your firmware.*

⑤ Click the Update the firmware and software for your Microsoft Broadband networking devices link.

A page that lets you download the firmware update appears.

⑥ Click the link for the latest firmware update.

---

**TIP**

**Did You Know?**
You can save downloads to your desktop, where you can easily find them. You can delete a file after you update your hardware. You can create a directory called Downloads, or something similar, to store the software you download.

continued

Before you actually download and install the firmware updates, take the time to read any release notes — information from the company about the updates — to see if any compatibility problems exist between your version of the hardware, your computer, and the new firmware updates. You may want to download and

install the firmware update to a test machine before deploying it to multiple computers.

Firmware is available on most hardware manufacturer Web sites. Updating firmware reprograms an actual microchip in the router.

### DOWNLOAD FIRMWARE
*(Continued)*

The File Download – Security Warning dialog box appears.

⑦ Click Save.

The Save As dialog box appears.

⑧ Select where you want to save the download.

⑨ Click Save.

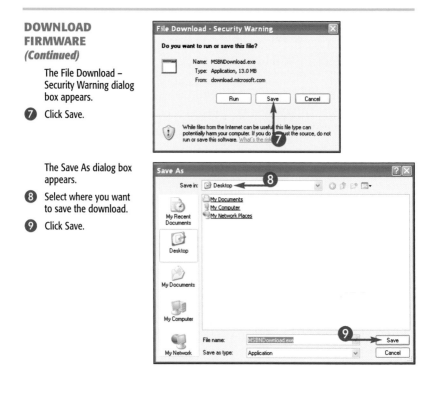

The window appears
showing you the
progress of the
download.

admin

4. On the **Management** menu, click **Back up and Restore Settings**.
5. Click **Back Up Settings**.
6. Click **Save**.
7. Type a name for the file that contains your base station settings or use the default name **Config.bin**.
8. Click the folder where you want to save the file, and then click **Save**.

· Back to the top

**Disable firewall or antivirus programs**

For information about how to disable your firewall or antivirus programs, view the documentation that came with the program, or contact the manufacturer's technical support.

· Back to the top

**Install firmware update**

**Warning** Do not install the firmware update from a computer that is connected to your base station by using a wireless network connection. If your wireless signal is interrupted, weakened, or interfered with while the firmware update is running, your base station may become unusable.

**Automatic update**

1. Start the Microsoft Broadband Network Utility.
2. On the **Help** menu, click **Check for Updates Online**. The Broadband Networking Update Service checks for updates by connecting to Microsoft. This can take several minutes, depending on your connection speed. If any new updates are available, a download screen appears. Otherwise, you receive a "No new updates are available" message.
3. Follow the instructions on the Broadband Networking Update Service page to update the firmware of your device.

If the automatic update fails, go to the "Manual Update" section.

**Manual update**

1. Download the MSBNDownload.exe firmware update. To do this, visit the following Microsoft Web site:

   http://go.microsoft.com/fwlink/?linkid=6790

2. Save the file to a location such as your desktop or the My Documents folder.
3. If you are updating a wireless base station, use an Ethernet cable to connect your computer to your base station.
4. To run the update, double-click the **MSBNDownload** file.
5. Follow the instructions that appear on the screen to update the firmware of your device.

While the firmware is being saved to your base station, the power light on the base station blinks and then turns orange. When the light is solid green, the update is completed. If the update fails, the power light continues to blink slowly until you successfully update the firmware. In this situation, you can try to update the firmware again, or you can reset the base station.

· Back to the top

After the firmware
downloads, you see the
download Web page.

● You can follow the
directions on this page
to download the
firmware you need.

## TIP

**Did You Know?**

Manufacturers often print their Web site addresses on packaging or in
instruction manuals. You also can use a search engine to locate the Web
address for a manufacturer.

After downloading the latest version of firmware, you need to upgrade your hardware. Upgrading the firmware applies updates that may improve the performance of the device or increase the security.

In some cases, you can send your downloaded file to the hardware using a Web browser. In other cases, you need to upgrade the firmware by running a setup routine provided by the firmware update. Usually this setup routine is part of the file you download. You need to consult the manufacturer of your router to find out how to upgrade the firmware you use.

### UPGRADE FIRMWARE

① In your Web browser, type the address of your access point and press Enter.

The access point home page appears.

② Click Management.

③ Click Upgrade Firmware.

The Upgrade Firmware page appears.

④ Click Browse.

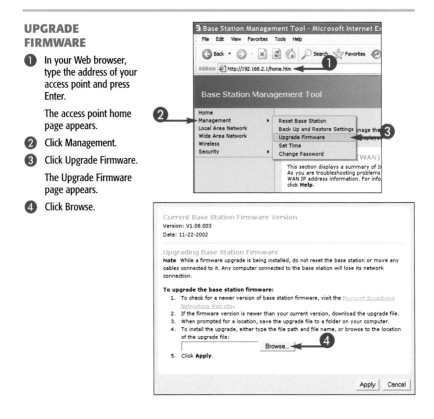

**Current Base Station Firmware Version**
Version: V1.08.003
Date: 11-22-2002

Upgrading Base Station Firmware
**Note** While a firmware upgrade is being installed, do not reset the base station or move any cables connected to it. Any computer connected to the base station will lose its network connection.

**To upgrade the base station firmware:**
1. To check for a newer version of base station firmware, visit the Microsoft Broadband Networking Web site.
2. If the firmware version is newer than your current version, download the upgrade file.
3. When prompted for a location, save the upgrade file to a folder on your computer.
4. To install the upgrade, either type the file path and file name, or browse to the location of the upgrade file:

   Browse...

5. Click **Apply**.

Apply    Cancel

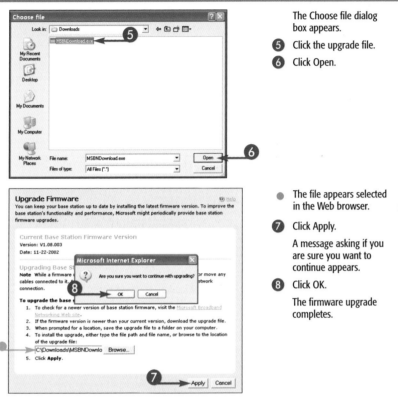

The Choose file dialog box appears.

**⑤** Click the upgrade file.

**⑥** Click Open.

● The file appears selected in the Web browser.

**⑦** Click Apply.

A message asking if you are sure you want to continue appears.

**⑧** Click OK.

The firmware upgrade completes.

---

## Did You Know?

You do not have to upgrade firmware. In most cases, hardware runs well on the firmware version the manufacturer installs. However, new firmware versions you download from manufacturer Web sites may add new features to your hardware. Upgrading also can sometimes improve the speed and quality of your wireless network.

# Understanding DHCP

Your wireless network router automatically assigns each of your computers a unique network address using *DHCP*, or *Dynamic Host Configuration Protocol*. DHCP makes it easier to add computers and other devices to your wireless network.

Microsoft Windows XP has built-in support for DHCP. When you first set up Windows, if you are already part of a network, you are prompted if you want to set up DHCP. If, however, you are not initially part of a network and you now want to attach to one that requires DHCP, you need to install and configure DHCP for your network.

## IP Address

An *IP*, or *Internet Protocol*, address identifies your computer in a unique way, similar to how your postal mail address identifies where you live. An IP address is 32 bits long and looks like this: 208.215.179.146.

## DHCP Server

Your wireless network gateway hardware probably has a built-in DHCP server. It assigns each computer on your network a dynamic IP address, either when a PC is turned on or when you manually request an IP address. In a computer-to-computer network, your host PC may act as a DHCP server, assigning IP addresses to client computers.

## DHCP Clients

Your Windows XP computers are configured by default to automatically request an IP address from a DHCP server. In some cases, such as when setting up a host computer for a computer-to-computer network, you should turn off this automatic feature and assign your computer a permanent, or static, IP address.

A static IP address is one that is permanently assigned to your computer. It does not change, even if you connect to a different wireless network. In cases in which you access multiple networks, you may need to re-enable DHCP for those networks that have the DHCP server.

## Network Address Translation

Your router likely has built-in Network Address Translation (NAT) support. NAT is an Internet standard that provides your internal network with its own set of IP addresses, usually beginning with 192.168.1.1, which do not conflict with Internet addresses. NAT also hides your computers so Internet users cannot see them from outside your network. Depending on the router and its configuration settings, NAT tables can become outdated and not include the most current addressing information. Administrators can force the router to update the NAT when several new computers join the network.

# Installing Wireless Hardware in PCs

Your computer communicates with your wireless network using a network adapter. You can install an internal or external adapter, depending on whether your computer is a desktop or laptop PC.

When you shop for wireless equipment, one of the first components to check on is the type of wireless networking card you need to use in your computer. Wireless network cards look a little different than wired network cards. In some cases, the wireless card has a small antenna attached to it.

In other cases, such as some wireless cards you insert into your laptop's PC card slot, the card does not have an antenna; instead, the end of the card that sticks out of the slot is thicker than the rest of the card. This is an internal antenna for the card.

To communicate with a wireless network, you may need to install hardware in your computer.

Some computers require two or more network adapters to communicate with multiple networks. As long as the computer has connections for the adapters, there is no restriction on the number of network adapters that you can attach or insert into a computer. You can also use wireless network adapters in the same computer that has standard Ethernet cable network adapters.

When you purchase a new computer, most retailers install and configure other computer devices at the same time. If you purchase a new computer with a wireless network adapter, you can avoid installing and configuring the adapter yourself by purchasing the equipment from the same retailer who will perform the installation.

## Software Drivers

The wireless network adapter includes software that you must install before you insert the wireless card. You may even have to install the software before installing the hardware. The software contains device drivers to enable the network card to communicate with the operating system. Some operating systems may already have the necessary drivers built in, and the installation of software is not necessary. Consult your network card's documentation for more information.

## Antenna

Most wireless adapters have an antenna attached for transmitting and receiving signals from other wireless network devices. The antenna is attached to the network card and placed at the rear of the computer. Position the computer and leave enough clearance to install and move the antenna, because moving the antenna may increase the signal quality.

### Lights

Most network adapters, including wireless network adapters, have a number of lights on the adapter that you can use to provide information about the status of the adapter. A power light indicates that the adapter is receiving power. A connection light may indicate the network adapter has detected and connected to a wireless network. When initially testing your wireless network adapter, you should position your computer and adapter so that you can view the lights on the wireless network adapter.

### Warranty

All computer hardware that you install in your computer should come with a warranty. The validity of the warranty of the hardware, and even the computer, may depend on you following the correct installation process for your card. For example, some computer manufacturers may void your warranty if you install your own hardware. Always refer to the warranty specifications for your computer and your new hardware for details on any steps you must take to avoid invalidating the warranty.

### Wireless Network Kits

Some manufacturers now sell all the necessary hardware to create a wireless network in a complete kit. The kit usually consists of one or two network adapters and a wireless network router. When purchasing a wireless network kit, you must ensure that the network adapter in the kit is suitable for use with your computer.

Each computer on your wireless network requires a wireless network adapter. Your network adapter acts as a radio transceiver, sending and receiving data across your wireless network. You can choose from several types of network adapters, which are also called *network interface cards*, or *NICs*.

You must connect a NIC either to or inside your computer. Some NICs require you to open the case of your computer and install the device internally. Others connect to one of your USB (Universal Serial Bus) ports from the outside. Another type slides into a laptop computer's PC card slot.

For some small handheld devices, you use CompactFlash wireless adapters.

## PCI Adapter

Most computers have expansion slots inside the computer that allow you to add functionality to a computer. You can install a network adapter in a *PCI*, or *Peripheral Component Interconnect*, slot inside your desktop computer. PCI slots are the most common type of expansion slots in a computer. You need to open the computer case, find an empty PCI slot, and then install the adapter.

## PC Card Adapter

You can install a *PC card-compatible network adapter* in your laptop computer. A PC card is a small credit-card-sized adapter that fits into most laptop computers, adding functionality to the laptop.

### USB Adapter

A *USB network adapter* plugs into a USB slot, which is available on both desktop and laptop computers. USB adapters are quick and easy to install. You do not have to open the computer to install them, and in many cases, the operating system automatically configures the USB.

### CompactFlash Adapter

Handheld computers are very small computers that perform a wide variety of information management tasks, such as tracking expenses, recording contact information, and scheduling meetings. Many handheld computers can connect to a wireless network using a *CompactFlash wireless adapter card*. Once connected to the wireless network, you can exchange information between the handheld computer and other computers on the network.

CompactFlash adapters incorporate security technology to provide 64- and 128-bit encryption to help ensure your data is secure as you access wireless networks. Some CompactFlash adapters also include a port for wired network connections, making it possible for your handheld device to connect to conventional wired networks. CompactFlash adapters are easy to use because all you do is slide them into your device and let your device configure them. Finally, these adapters do not require a separate power source to run them.

You can install a PCI-compatible network adapter in a desktop computer or a PC card-compatible network adapter in a laptop or notebook computer so your computer can communicate with your wireless network.

Installation steps depend on the adapter. Although the actual steps may vary, in most cases you use wizards, so the steps in this section are similar to what you see.

Consult your network card instructions for specific procedures.

Typically, you install a network adapter in a PCI, or *Peripheral Component Interconnect*, by opening the computer case, finding an empty PCI slot, and then installing the adapter. To install a PC card in your notebook, you simply insert the card into the side of your laptop computer.

## INSTALL A NETWORK ADAPTER

1 Insert the installation CD-ROM.

The wireless adapter installation screen appears.

2 Click Next to continue.

The Before continuing with Setup screen appears.

3 Click Next.

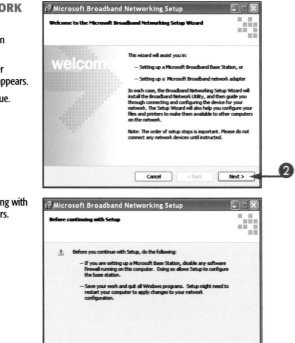

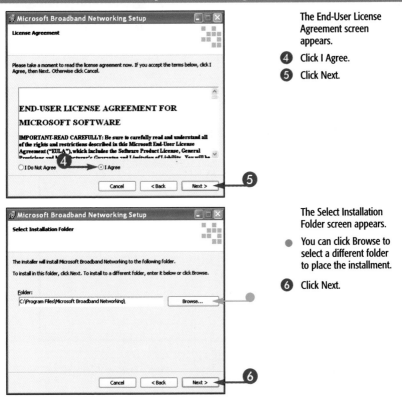

The End-User License Agreement screen appears.

④ Click I Agree.

⑤ Click Next.

The Select Installation Folder screen appears.

● You can click Browse to select a different folder to place the installment.

⑥ Click Next.

**TIP**

**Did You Know?**
You can install the software on the CD-ROM and then upgrade it to the latest version from the manufacturer's Web site after your wireless network is up and running.

continued

You can configure your wireless network adapter with custom software that many manufacturers supply with their hardware. Some of the utilities provided enable you to modify configuration settings on the adapter, such as speed and protocol settings. Others give you the capability to monitor and log adapter performance. With the monitoring and logging capabilities, you can track and troubleshoot problems that may arise with your network adapter.

You should disable Automatic Wireless Network Configuration if you are installing and using configuration utility software.

The network adapter utility that installs with your network adapter may be different than the one shown here.

The Software Installation is Complete screen appears.

 Click Next to finish.

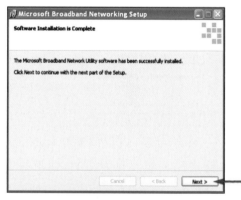

 Plug in the USB wireless adapter to a USB port on your computer.

● A Wireless Network Connection icon appears on the taskbar.

## USING NETWORK ADAPTER UTILITIES

① Double-click the Broadband Network Utility icon on the taskbar.

**Note:** *Windows XP may warn you that the software has not passed Windows logo testing; if so, click Continue.*

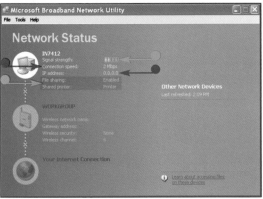

The Microsoft Broadband Network Utility dialog box appears.

● You can view the signal strength.

● You can view the connection speed.

● You can view the IP address.

● You can view file and printer sharing status.

### Did You Know?

Microsoft tests many devices that will work with Windows. Devices that Microsoft has not certified generate the logo warning error when you install them. Many devices that generate the logo warning were created after Windows was last updated, or they are new and may be certified in the future. For more on the wireless networking features and capabilities, click the Learn about accessing files on these devices link on the Microsoft Broadband Network Utility dialog box.

You can confirm the successful installation of your network adapter in Windows XP Device Manager. You can also reinstall the software driver for your hardware if Device Manager does not list your network adapter.

The Device Manager is a helpful feature to use when you are experiencing problems with your wireless networking connection. With the Device Manager, it is easy to see the hardware devices installed on your computer and the status of the hardware — working or not working.

The Device Manager also enables you to remove a hardware device or reinstall a device that may not be working correctly.

① Click start.

② Click Control Panel.

The Control Panel appears.

③ Click Performance and Maintenance.

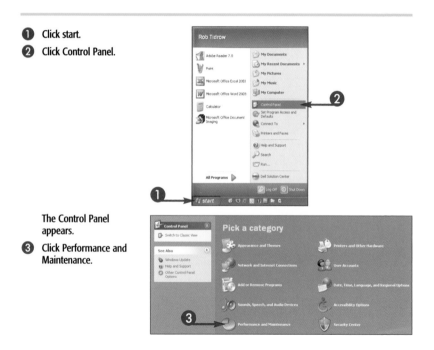

The Performance and Maintenance window appears.

④ Click the System icon.

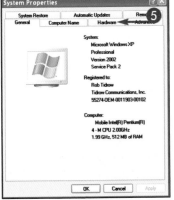

The System Properties dialog box appears.

⑤ Click the Hardware tab.

---

**TIP**

**Did You Know?**

You may need to confirm driver installation if you are experiencing difficulty with your network hardware. Incorrect driver installation can cause many different types of problems with network adapters.

continued

Many manufacturers regularly update and release new software drivers for your hardware, providing new features and better stability. You can download these updates and install them on your computer.

As with any inserted device, your new network adapter may attempt to use resources already allocated to another device in the computer. Check your operating system's documentation to resolve resource conflicts, or to

temporarily remove any unnecessary hardware, such as a sound card, until you configure the wireless network adapter and have it working.

If the network adapter appears correctly installed, but you cannot communicate with an existing network, double-check the network adapter settings. A simple typing error during network identification can prevent your network adapter from communicating with any wireless networks.

**⑥** Click Device Manager.

The Device Manager window appears.

**⑦** Click the plus sign beside Network adapters.

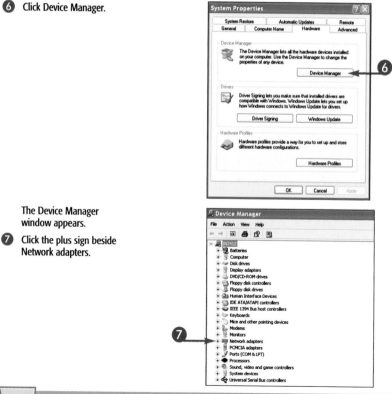

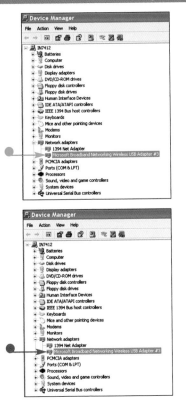

A list of network adapters appears.

● The icon beside the wireless PC card shows the driver is installed and the adapter is working.

● The red X through the icon shows the network adapter is not working properly.

## TIP

### Did You Know?

To uninstall or disable a network adapter driver, right-click a network adapter in the Device Manager. Click Disable or Uninstall. Your computer disables or uninstalls the network adapter driver.

Most manufacturers of network adapters provide updates to their hardware. These updates are called firmware updates. Periodically, you may want to visit your adapter manufacturer's Web site to learn about firmware updates that may be available.

If you encounter problems when installing or using the adapter, you also should visit this Web site. Find out and download any

updates available for your adapter. Some manufacturers also include forms you can fill out to receive e-mails or other types of correspondence alerting you to new updates and releases.

Finally, after you download any updates to your firmware, you must install them on your computer. This activates them for your wireless network adapter.

① In your Web browser, type **http://support. microsoft.com/kb/ 814445** and press Enter.

**Note:** If your hardware is made by a different manufacturer, type the Web address of the manufacturer that makes your hardware.

The Microsoft page for updating the adapter firmware appears.

② Click the filename.

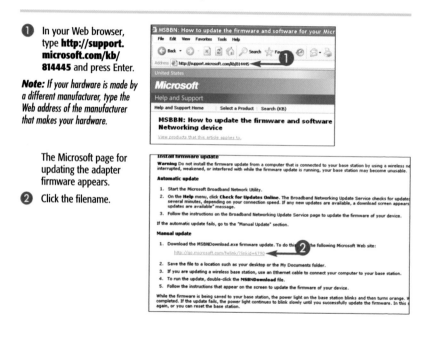

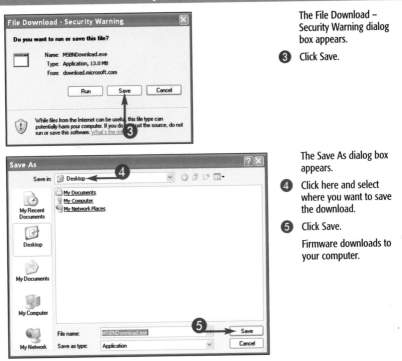

The File Download – Security Warning dialog box appears.

③ Click Save.

The Save As dialog box appears.

④ Click here and select where you want to save the download.

⑤ Click Save.

Firmware downloads to your computer.

**TIP**

**Did You Know**

After you download the firmware update to your computer, locate the file using My Computer or Windows Explorer and double-click the file. This activates the setup program, which then walks you through installing the update on your computer.

# Configuring Wireless Networks

After you install a wireless network adapter in your computer, you can configure your wireless networks and view how well they are working.

One point to keep in mind with Windows XP computers using wireless technologies is that they can connect to multiple wireless networks, but can connect to only one network at a time. If you use one particular

wireless network, you should ensure that you have specified the correct settings – called the *service set identifier* or SSID – of the network you want to connect to, otherwise Windows XP may connect to another wireless network instead.

In this chapter, you learn how to configure and monitor networks.

# Quick Tips

# View Available Networks

Windows XP can automatically configure your wireless network adapter so that you can view available networks. You can save time by using this feature rather than manually configuring your wireless network.

When you turn on your computer and have a wireless network adapter installed, Windows searches the area for wireless signals. Windows picks up radio and

infrared signals from your surroundings to find these networks. These wireless networks are then displayed in a list of available networks.

A nice feature of Windows XP is the taskbar messages that display when it finds a network. In the lower right corner of the taskbar by the clock, a small window appears when an available network is found.

① Right-click the Network Connection icon.

② Click View Available Wireless Networks.

The Wireless Network Connection dialog box appears.

③ Click the network to which you want to connect.

④ Click Connect.

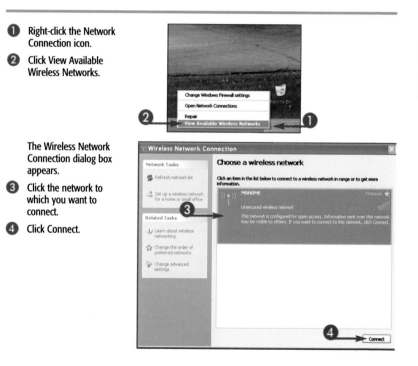

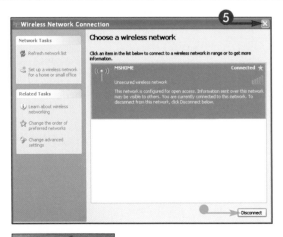

- The Connect button changes to the Disconnect button after your computer connects to the wireless network.

⑤ Click the Close button to exit from the Wireless Network Connection dialog box.

- A text balloon appears, showing the speed and signal strength of your wireless network.

*Note:* For more about network speed and signal strength, see the section "View Signal Strength."

**Did You Know?**
The Wireless Network Connection dialog box displays all wireless networks within reception range of your computer. If it displays more than one network, either there are multiple wireless networks available in your home or office, or you are picking up signals from an open network operating in a neighboring business or residence.

# Configure an Available Network

Even with the automatic configuration of wireless networks, some wireless networks may not appear in the list Windows finds. Most wireless networks used by businesses, schools, and hospitals require you to configure a network connection manually to help keep out unauthorized users. In these cases, you must manually configure the network.

You can configure an available wireless network. Your computer saves your settings and adds the network to the Preferred networks list. To make changes to configure a network, you must log on using an administrator's account.

① From the Wireless Network Connection dialog box, click Change advanced settings.

*Note:* To access the Wireless Network Connection dialog box, see the section "View Available Networks."

*Note:* You must have administrator privileges on your computer to configure settings in the Wireless Network Connection dialog box.

The Wireless Network Connection Properties dialog box appears.

② Click the Wireless Networks tab.

③ Select a network from the Preferred networks list.

④ Click Properties.

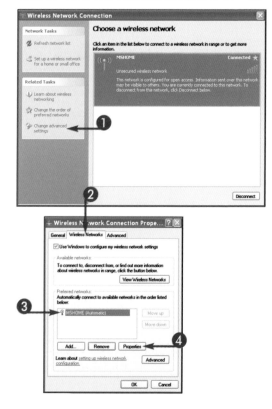

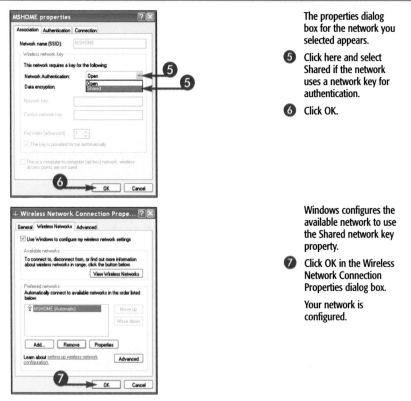

The properties dialog box for the network you selected appears.

⑤ Click here and select Shared if the network uses a network key for authentication.

⑥ Click OK.

Windows configures the available network to use the Shared network key property.

⑦ Click OK in the Wireless Network Connection Properties dialog box.

Your network is configured.

---

**TIP**

**Did You Know?**

You can manually update the list of available networks. To do this, on the Wireless Networks tab of the Wireless Network Connection Properties dialog box, click the Refresh button in the Available Networks area.

You can add a preferred network to your list of available networks. Windows XP automatically connects to networks that appear in this list.

A *preferred network* is one that Windows attempts to connect to each time you turn on your computer and you have an installed wireless network adapter. If a wireless network in your Preferred networks list is not available or if the network is out of range, Windows displays a red X next to the network name to show that it is unavailable.

To add a network to the preferred list, you need to know the name of the network — its SSID, or service set identifier — and if it requires a network key to connect to it.

---

① From the Wireless Network Connection Properties dialog box, click Add.

**Note:** *To access the Wireless Network Connection dialog box, see the section "View Available Networks."*

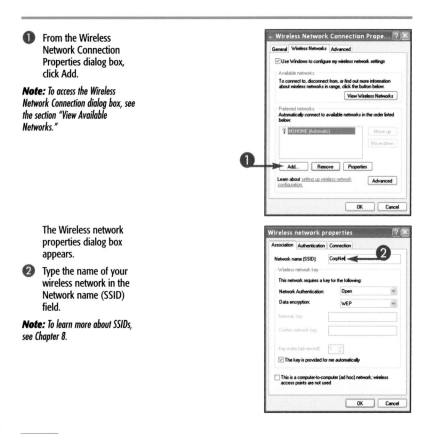

The Wireless network properties dialog box appears.

② Type the name of your wireless network in the Network name (SSID) field.

**Note:** *To learn more about SSIDs, see Chapter 8.*

**Wireless network properties**

Association | Authentication | Connection

Network name (SSID): CorpNet

Wireless network key

This network requires a key for the following:

Network Authentication: Open

Data encryption: WEP

Network key:

Confirm network key:

Key index (advanced): 1

☑ The key is provided for me automatically

☐ This is a computer-to-computer (ad hoc) network; wireless access points are not used

**③** → OK | Cancel

**③** Click OK.

**Wireless Network Connection Prope...**

General | Wireless Networks | Advanced

☑ Use Windows to configure my wireless network settings

Available networks:

To connect to, disconnect from, or find out more information about wireless networks in range, click the button below.

View Wireless Networks

Preferred networks:
Automatically connect to available networks in the order listed below:

CorpNet (Automatic)    Move up

MSHOME (Automatic)    Move down

Add... | Remove | Properties

Learn about setting up wireless network configuration.    Advanced

OK | Cancel

● Windows XP adds the network to the Preferred networks list.

**TIP**

**Did You Know?**

You can configure the Wireless network key settings in the Association tab of the Wireless network properties dialog box. These settings must match the network key settings of the network you are adding. To learn about wireless network security, see Chapter 8.

# Remove a Preferred Network

You can remove a wireless network from the Preferred networks list when you no longer need to frequently connect to the network. You may want to do this after you leave an area in which a public wireless network was available and you do not plan to visit there soon. Or you may no longer need to access a company-wide wireless network using your personal laptop.

By removing the network from the list, Windows XP no longer looks for this network when you start your computer. This speeds up Windows' boot time, because Windows no longer searches for (and connects to if it is found) a wireless network you do not want to access.

① From the Wireless Network Connection Properties dialog box, click a preferred network.

***Note:*** *To access the Wireless Network Connection dialog box, see the section "View Available Networks."*

② Click Remove.

Windows removes the preferred network.

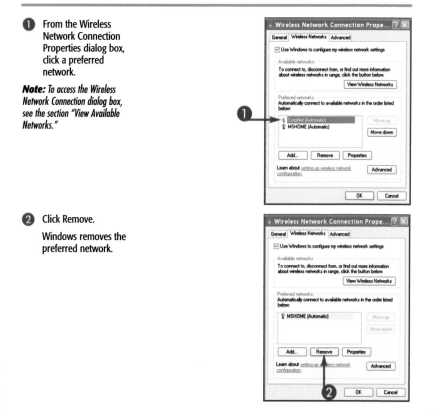

You can rearrange available networks in the Preferred networks list. Because Windows XP uses the list to determine the order in which it connects to your preferred networks, you move a network to the bottom of the list if you know one network is not available in your area. Windows finds another network first and does not search for the unavailable network.

Also, you may want to place a preferred network at the top of the list if you know it is the one you most frequently want to log on to, but you want to keep other wireless networks listed for future usage. You do not need to reconfigure the other networks even though you use them less frequently than your main network.

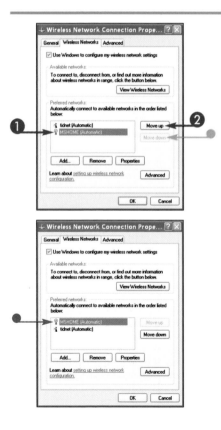

① From the Wireless Network Connection Properties dialog box, click a preferred network.

*Note: To access the Wireless Network Connection dialog box, see the section "View Available Networks."*

② Click Move up.

● You can click Move down to move a network lower on the list.

● The preferred network moves higher on the Preferred networks list.

You can view a signal strength meter in Windows XP that displays the strength of the radio waves your computer receives from your wireless network. The signal strength meter helps you determine if you are within range of your wireless network, and how fast information is moving across your network.

The strength of a wireless network signal depends on several factors. Among these factors are the distance from the wireless network, the placement of your computer next to other devices that can interfere with signals, and your computer's battery life.

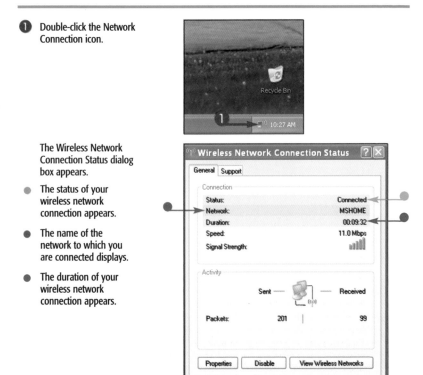

① Double-click the Network Connection icon.

The Wireless Network Connection Status dialog box appears.

● The status of your wireless network connection appears.

● The name of the network to which you are connected displays.

● The duration of your wireless network connection appears.

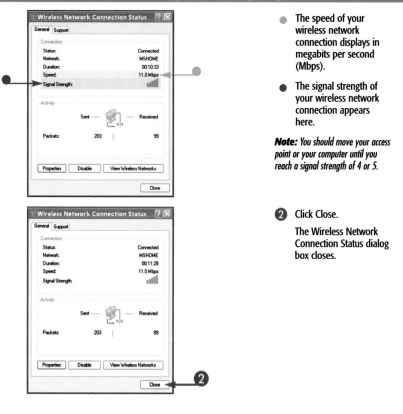

- The speed of your wireless network connection displays in megabits per second (Mbps).

- The signal strength of your wireless network connection appears here.

*Note: You should move your access point or your computer until you reach a signal strength of 4 or 5.*

② Click Close.

The Wireless Network Connection Status dialog box closes.

## Did You Know?

There are other ways you can view signal strength. You can position your mouse pointer over the Network icon in your taskbar notification area. A text balloon appears, displaying network signal strength and speed. You also can view your signal strength with software utilities made by your hardware manufacturer.

Windows XP helps you quickly create a wireless network with the Wireless Zero Configuration feature. When you confirm that the Wireless Zero Configuration feature is on, you can avoid problems when you install and configure your hardware.

Usually, Windows XP automatically turns on the Zero Configuration service when you install a wireless networking device, such as a wireless network interface adapter. However, there are times when Windows XP does not automatically enable the service or when another device disables the service. In either case, you should enable it to help you configure your wireless network connection.

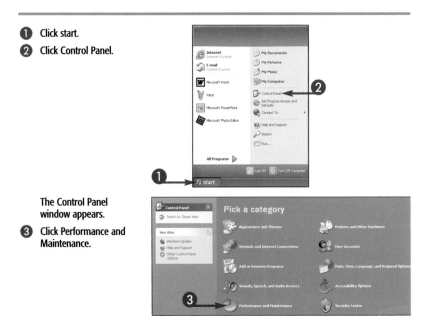

① Click start.

② Click Control Panel.

The Control Panel window appears.

③ Click Performance and Maintenance.

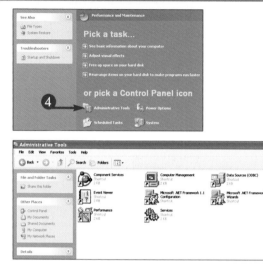

The Performance and Maintenance window appears.

④ Click Administrative Tools.

The Administrative Tools window appears.

## TIP

### Did You Know?

You can set up a wireless network interface card if Windows XP fails to configure it automatically. First, confirm that the Windows XP Wireless Zero Configuration feature is enabled. To confirm Zero Configuration is on, perform Steps **1** to **7**. If this feature is enabled but Windows XP does not configure your hardware, you can install configuration software that most manufacturers include with their network adapters.

continued

# Confirm Zero Configuration
## Is On *(continued)*

You can use Automatic Wireless Network Configuration with newer wireless network interface cards that Microsoft certifies as working with Windows XP.

Some older wireless network interface cards (NICs) may have updated software drivers you can download from the NIC manufacturer's Web site.

You need administrator privileges to turn Wireless Zero Configuration on or off. If you do not have administrator privileges, contact your system administrator or IT (information technology) department to ask how you can start the Wireless Zero Configuration service on your computer.

⑤ Double-click the Services icon.

The Services window appears.

⑥ Scroll down to the bottom of the window.

⑦ Click Wireless Zero Configuration.

⑧ Click the Start the service link.

**Note:** If "Stop the service" appears instead of "Start the service," your computer is already running the Wireless Zero Configuration service.

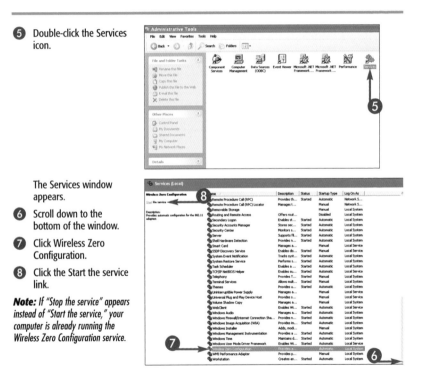

| Uninterruptible Power Supply | Manages a... | | Manual | Local Service |
|---|---|---|---|---|
| Universal Plug and Play Device Host | Provides s... | | Manual | Local Service |
| Volume Shadow Copy | Manages a... | | Manual | Local System |
| WebClient | Enables Wi... | Started | Automatic | Local Service |
| Windows Audio | Manages a... | Started | Automatic | Local System |
| Windows Firewall/Internet Connection Sha... | Provides n... | Sta... | Automatic | Local System |
| Windows Image Acquisition (WIA) | Provides im... | Started | Automatic | Local System |
| Windows Installer | Adds, modi... | | Manual | Local System |
| Windows Management Instrumentation | Provides a ... | Started | Automatic | Local System |
| Windows Time | Maintains d... | Started | Automatic | Local System |
| Windows User Mode Driver Framework | Enables Wi... | Sta...ed | Automatic | Local Service |
| Wireless Zero Configuration | Provides a... | Started | Automatic | Local Service |
| WMI Performance Adapter | Provides p... | | Manual | Local System |
| Workstation | Creates an... | Started | Automatic | Local System |

- The Wireless Zero Configuration service starts.

**9** Click the Close button.

The Services window closes.

| Name | Description | Status | Startup Type | Log On As |
|---|---|---|---|---|
| Remote Procedure Call (RPC) | Provides th... | Started | Automatic | Network S... |
| Remote Procedure Call (RPC) Locator | Manages t... | | Manual | Network S... |
| Removable Storage | | | Manual | Local System |
| Routing and Remote Access | Offers rout... | | Disabled | Local System |
| Secondary Logon | Enables st... | Started | Automatic | Local System |
| Security Accounts Manager | Stores sec... | Started | Automatic | Local System |
| Security Center | Monitors s... | Started | Automatic | Local System |
| Server | Supports fil... | Started | Automatic | Local System |
| Shell Hardware Detection | Provides n... | Started | Automatic | Local System |
| Smart Card | Manages a... | | Manual | Local Service |
| SSDP Discovery Service | Enables dis... | Started | Manual | Local Service |

**Did You Know?**

If the service is already running, you can restart the Wireless Zero Configuration service. This may be helpful when you need to troubleshoot your wireless network. To restart the service, perform Steps **1** to **7**. In step **8**, click Restart the service.

# Disable and Enable Automatic Configuration

Automatic Wireless Network Configuration automatically configures your wireless network adapter and connects you to your wireless network. You can disable the Automatic Wireless Network Configuration feature in Windows XP when you need to install software updates that the manufacturers provide with their network interface cards.

After you install software updates, you need to enable the Automatic Wireless

Network Configuration to let Windows automatically configure your wireless network adapter each time you boot your computer.

If the Automatic Wireless Network Configuration feature is disabled, the next time you restart Windows it may reenable automatically. If it does not, you have to reenable it manually using the steps here.

## DISABLE AUTOMATIC CONFIGURATION

① From the Wireless Network Connection dialog box, click to uncheck the Use Windows to configure my wireless network settings option.

*Note:* To access the Wireless Network Connection dialog box, see the section "View Available Networks."

② Click OK.

Windows disables the automatic configuration of your wireless network settings.

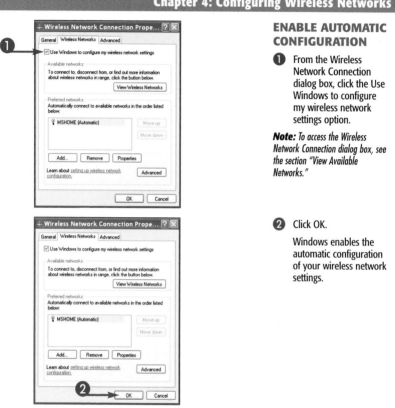

**ENABLE AUTOMATIC CONFIGURATION**

**①** From the Wireless Network Connection dialog box, click the Use Windows to configure my wireless network settings option.

*Note: To access the Wireless Network Connection dialog box, see the section "View Available Networks."*

**②** Click OK.

Windows enables the automatic configuration of your wireless network settings.

**Did You Know?**

When using a laptop or portable computer, you should keep automatic configuration turned on as a default. This way Windows will be able to connect to and configure your wireless network settings when you enter an area that has a wireless network available.

# Create a Wireless Bridge

You can connect, or *bridge*, one network with another network. For example, you can bridge an Ethernet-wired network with your wireless network. A network bridge lets you connect your laptop computer to your home or office network without requiring an additional wireless access point.

Another place where you might see a wireless network bridge is with gaming console setups. You can use a bridge to connect your game console (a Microsoft Xbox, for example) to your wireless network and other computers.

① Right-click the Network Connection icon.

② Click Open Network Connections.

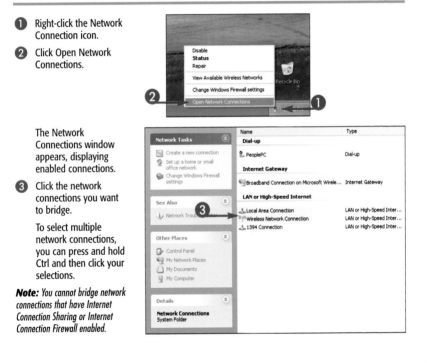

The Network Connections window appears, displaying enabled connections.

③ Click the network connections you want to bridge.

To select multiple network connections, you can press and hold Ctrl and then click your selections.

**Note:** *You cannot bridge network connections that have Internet Connection Sharing or Internet Connection Firewall enabled.*

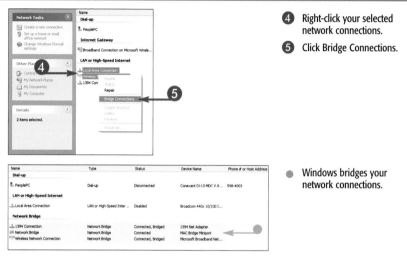

④ Right-click your selected network connections.

⑤ Click Bridge Connections.

● Windows bridges your network connections.

**Did You Know?**

A computer that is bridging two or more networks needs at least two network adapters. In the case of an Ethernet-wired network and a wireless network, your computer needs an Ethernet adapter and a wireless network adapter.

# Creating Computer-To-Computer Networks

Computers can communicate directly with each other through a computer-to-computer network. In this chapter, you learn how to create a computer-to-computer network and share an Internet connection.

One reason to set up a computer-to-computer network is cost. It is less expensive to set up than an infrastructure network because there are fewer pieces of equipment to purchase. You do not, for example, need to purchase a separate router to allow your computers to network with each other.

# Understanding Computer-to-Computer Considerations

Many issues can help you decide if you should use a computer-to-computer network, or a more centralized network setup. Some of these include overall cost of the network, features of the operating systems, how well the computers communicate with each other, the distance between the computers, and the practical size of the network.

## Cost

Connecting computers together in a computer-to-computer network requires each computer to have a wireless network adapter. Items, such as a wireless router or an access point, are not required. This makes a computer-to-computer network less expensive to set up than a more centralized network.

## Operating Systems

Most modern operating systems have enough network functionality built into them so that you can use network features between just two computers. Usernames, passwords, and file and print sharing are available on almost all operating systems, allowing you to take full advantage of the services and resources of all computers in the network.

## Compatibility

As long as the wireless adapters in each computer use the same network standard, the network adapters can communicate with each other regardless of the type or manufacturer of the wireless adapter. Internal wireless adapters installed inside a computer can communicate with other internal and external wireless adapters that you may attach to a computer USB port.

## Size

There is no practical limit to the number of computers that can communicate in a computer-to-computer network. As long as each computer has a wireless adapter and is in range of all the other computers in the network, they work on the network. In practice, if your network has more than five computers, you may want to consider using a more centralized network layout using a wireless switch or other type of wireless access point.

## Range

The distance between computers on a computer-to-computer network depends solely on the coverage area of the network adapters. You must have all network computers in range of each other for the network to work effectively. This means you can reduce the geographic size of the network when using a network adapter with a reduced range.

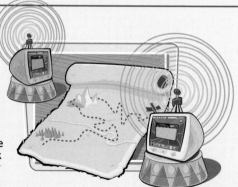

# Computer-to-Computer versus Infrastructure Networks

Each network type has its own benefits and disadvantages. Before setting up a larger and more costly infrastructure network, consider setting up a computer-to-computer network first. Then try out the capabilities of the network. If the capabilities and performance of this type of network are all you need, keep it.

However, if you need a network that requires centralized management, advanced network security, and backup resources for the network, then consider a larger, infrastructure network.

## Setup Speed

The time you need to set up a computer-to-computer network is significantly less than setting up an infrastructure network because less hardware is required. Once you install the wireless adapter in a computer and configure it, the network is operational and ready for use.

## Extra Equipment

Complexity and costs are reduced in a computer-to-computer network because you use fewer hardware devices. Administration is also easy to manage because each user is responsible for his or her own computer, and administrators are not required for devices, such as routers and access points.

### Easy to Change

Because a computer-to-computer network is not a fixed collection of wireless devices, you can add or remove wireless nodes as necessary.

### Redundancy

In computer-to-computer networks, multiple wireless devices communicate with each other. If one wireless device is no longer available, a computer can still communicate with the network by communicating with other devices in range. On an infrastructure network, if a single device, such as an access point fails, many users can be disconnected from the network.

### Mobile

Computer-to-computer networks are easy to move because the wireless devices are not permanent like wireless hubs. It is even possible to operate a computer-to-computer network between mobile computers located in moving vehicles.

You create a computer-to-computer wireless network with Internet access by connecting a second or third computer to a PC that has an Internet connection. You must configure one computer as the host PC.

The host PC is the "main" computer on the network for other client PCs to connect to. The host computer connects directly to the Internet connection, and then client computers connect through the host to the Internet.

① **Double-click the Network Connection icon.**

The Wireless Network Connection Status dialog box appears.

② **Click Properties.**

The Wireless Network Connection Properties dialog box appears.

③ **Click the Internet Protocol (TCP/IP) option.**

④ **Click Properties.**

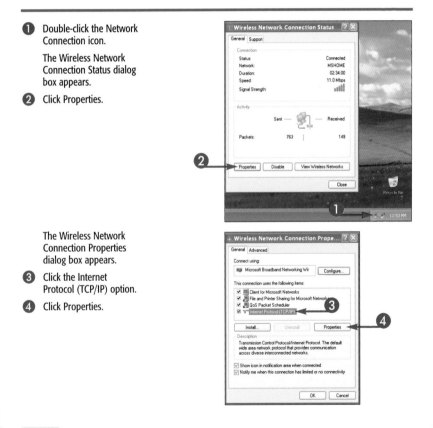

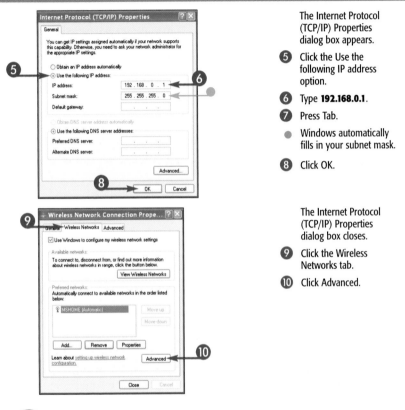

The Internet Protocol (TCP/IP) Properties dialog box appears.

⑤ Click the Use the following IP address option.

⑥ Type **192.168.0.1**.

⑦ Press Tab.

● Windows automatically fills in your subnet mask.

⑧ Click OK.

The Internet Protocol (TCP/IP) Properties dialog box closes.

⑨ Click the Wireless Networks tab.

⑩ Click Advanced.

**Did You Know?**

Computer-to-computer wireless networks are called by other names. You may hear the terms *peer-to-peer* or *ad hoc* networks to refer to these types of networks.

Your host PC needs either a modem or an Ethernet adapter for Internet access and a wireless adapter card. If you share a dial-up Internet access with other computers in a computer-to-computer network, note that the connection speed becomes slower because each computer uses part of the available bandwidth.

For this reason, you may want to invest in a high-speed connection to the Internet, such as broadband, when you decide to set up a shared Internet connection.

The Advanced dialog box appears.

⑪ Click the Computer-to-computer (ad hoc) networks only option.

⑫ Click to uncheck the Automatically connect to non-preferred networks option.

⑬ Click Close.

The Advanced window closes.

⑭ Click Add.

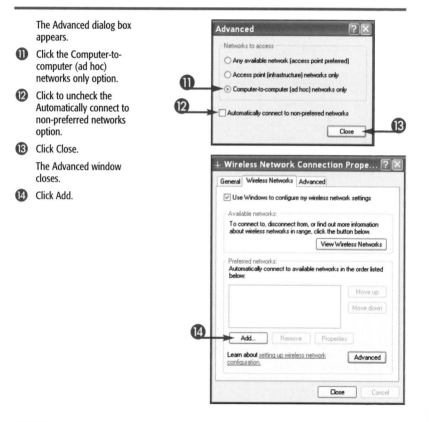

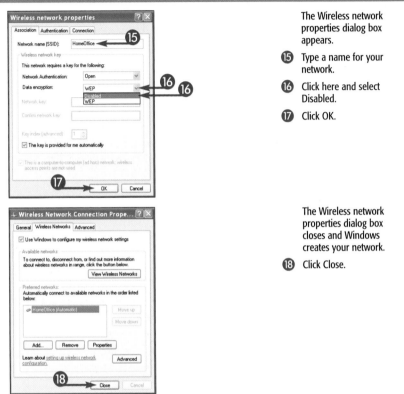

The Wireless network properties dialog box appears.

⑮ Type a name for your network.

⑯ Click here and select Disabled.

⑰ Click OK.

The Wireless network properties dialog box closes and Windows creates your network.

⑱ Click Close.

#### Did You Know?

If you turn off data encryption when you create your network, this makes it easier to configure your network. You should secure your wireless network once you configure it and confirm the client computers are communicating with your host computer. For more information on wireless network security, see Chapter 8.

You must configure the client computers on your wireless network. Your client PCs connect to the Internet through the host computer.

Even without Internet access, your computers can share files and folders,

printers, and other resources over a computer-to-computer wireless network. A wireless network without Internet access is more secure because unauthorized users cannot access your network from the Internet.

---

① From the Wireless Networking Connection Properties dialog box, click the Internet Protocol (TCP/IP) option.

*Note: To open the Wireless Network Connection Properties dialog box, see the section "Configure Host PC."*

② Click Properties.

The Internet Protocol (TCP/IP) Properties dialog box appears.

③ Click the Obtain an IP address automatically option.

*Note: If this client PC will share an Internet connection through your host computer, you can skip to Step 7.*

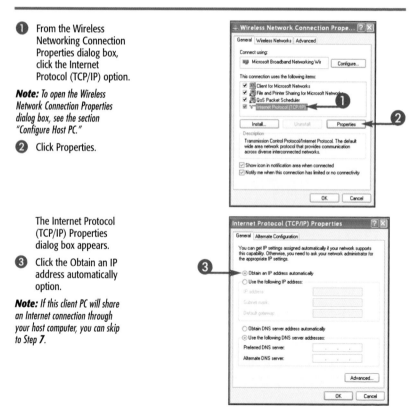

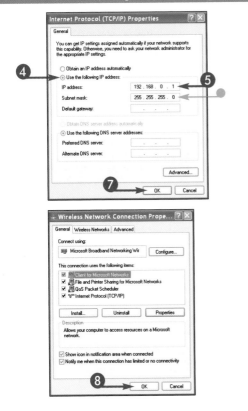

4 Click the Use the following IP address option.

5 Type a unique IP address for each client PC.

*Note:* You can start by typing **192.168.0.2**, increasing the last number for each client PC, for example, **192.168.0.3**, **192.168.0.4**, and so on.

6 Press Tab.

● Windows automatically fills in the Subnet mask number.

7 Click OK.

The Internet Protocol (TCP/IP) Properties dialog box closes.

8 Click OK.

9 Repeat Steps **9** to **18** in the section "Configure Host PC" to create a computer-to-computer network with the same name on each client PC.

Your client PC is configured for computer-to-computer networking.

**Did You Know?**

You can configure wireless network adapters for two modes, computer-to-computer and infrastructure mode. If the network adapter is configured to use infrastructure mode, then you can make it communicate with a computer-to-computer network.

# Enable Internet Sharing

You can share an Internet connection with other users on your computer-to-computer wireless network. You must first enable Internet sharing on your host PC.

One concern with sharing Internet connections is security. You can install third-party firewall software that lets you control the information moving both in and out of your computer. The Windows XP firewall only protects against incoming Internet connections.

① Double-click the Network icon.

② From the Local Area Connection Status dialog box, click Properties.

**Note:** *You should configure the wired network connection on your host PC, not its wireless network connection.*

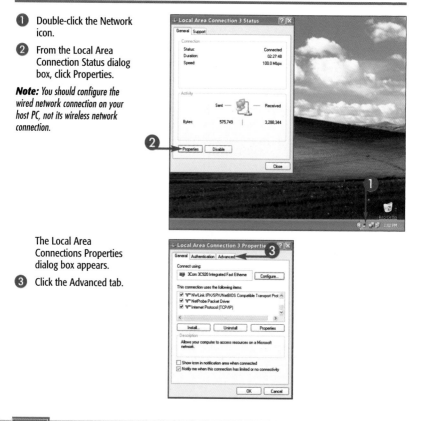

The Local Area Connections Properties dialog box appears.

③ Click the Advanced tab.

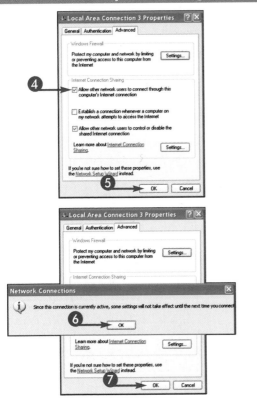

The Advanced tab appears.

④ Click the Allow other network users to connect through this computer's Internet connection option.

⑤ Click OK.

The Network Connections message box appears telling you that some settings may not take effect until the next time you connect to the Internet.

⑥ Click OK.

⑦ Click OK in the Advanced Setting dialog box.

⑧ Click OK in the Local Area Connection Status dialog box.

Windows enables Internet Connection Sharing.

**Did You Know?**

You cannot restrict computers that access the Internet through your host computer. Any client PC that is authorized to access your host computer over your computer-to-computer network shares the Internet connection.

# Working on Wireless Networks

The main reason for creating, connecting to, and administering a wireless network connection is to use it to get your work done. You can use wireless networks to share files, store and retrieve files in remote folders, and access devices in remote locations. In larger networks, you can even set up electronic mail servers to help users communicate with one another.

In this chapter, you learn how to browse your network, share a folder with other users on your network, and select a network printer.

*Quick Tips*

Most network users want to work with a number of common network services like printers, applications, and sharing and accessing files. This section describes common network services.

## Printers

Sharing relatively expensive equipment such as printers is one of the primary reasons that networks are useful. Using wireless networks makes it easier to share network printers because it is not necessary to attach a printer and computer using a common cable system.

## Share Files

You can easily use a wireless network to share the files on your computer with other users on the network. You can also allow other users to work on files stored on your computer.

### Access Files

You can store files that you access on a network on another user's computer or on a computer that is dedicated to storing files for all the users on the network. This type of computer is called a file server. Accessing file servers on a network often involves the use of authentication schemes such as user names and passwords. Storing files on a server provides several benefits, including enabling multiple users to modify a single document, allowing centralized backup support, and providing a central warehouse for your important files.

### Applications

You can store some applications on a computer network from which you can access and run the program. These computers, called *application servers*, usually control larger, more complex tasks, such as running a database program and managing the data in the database. Other types of applications that can run on application servers include document warehouses and multimedia servers. The latter type, for example, lets you store and view digital movies from one location. With document warehouses, a single program can store all the documents for a company and allow those documents to be edited and viewed from that storage environment.

You can connect to your wireless network by using a few different methods. One way is to boot up Windows XP so that it automatically searches for and connects to a preferred network.

Another way is to connect to the network using a shortcut in the Windows XP start menu. This is called *forcing a connection*. You may need to connect or reconnect to a wireless network after you install new hardware or when your connection is disabled or interrupted.

① Click start.

② Click Connect To.

③ Click Wireless Network Connection.

The Wireless Network Connection Status dialog box appears.

④ Click Close.

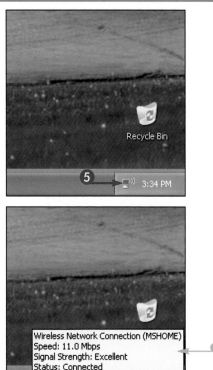

⑤ Position the mouse cursor over the Wireless Network Connection icon in the system tray.

● A text box appears, indicating the computer is connected to the wireless network.

Wireless Network Connection (MSHOME)
Speed: 11.0 Mbps
Signal Strength: Excellent
Status: Connected

**TIP**

**Did You Know?**
You can view all of your network connections by clicking start, Connect To, and then Show all connections. A Network Connections window appears, showing all your network connections.

You can use My Network Places to view the computers connected to your wireless network. My Network Places can appear in several places within Windows XP. You can open My Network Places from an icon on your desktop, from an icon on the Start menu, and from a folder in My Computer (or Windows Explorer).

My Network Places displays drives, folders, and files available on the network. You can access files, folders, and other information that is available to you on the wireless network.

① Click start.

② Click My Network Places.

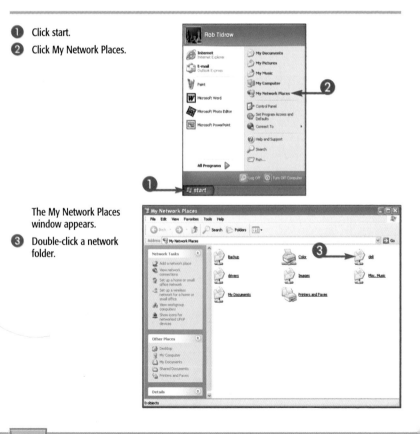

The My Network Places window appears.

③ Double-click a network folder.

The shared network folder opens displaying its contents.

④ Double-click a subfolder.

● The subfolder opens, displaying its content.

● You can click the Close button to close the window.

## Did You Know?

If you attempt to access a file or folder and it is not available, the computer that stores the information may have been removed from the network or is simply turned off. Check to see if the computer that is sharing the file or folder is turned on and connected to your wireless network.

You can share your files and folders with other users on your wireless network. This allows other users to access information on your computer that you choose to share.

When you share a folder on your wireless network, other users can view all the files within that folder. For this reason, share only those folders you want others to access. You should not share personal, confidential, and sensitive files with other users. Put those kinds of files in nonshared folders.

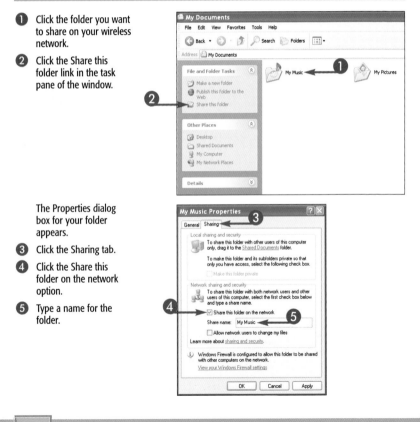

1 Click the folder you want to share on your wireless network.

2 Click the Share this folder link in the task pane of the window.

The Properties dialog box for your folder appears.

3 Click the Sharing tab.

4 Click the Share this folder on the network option.

5 Type a name for the folder.

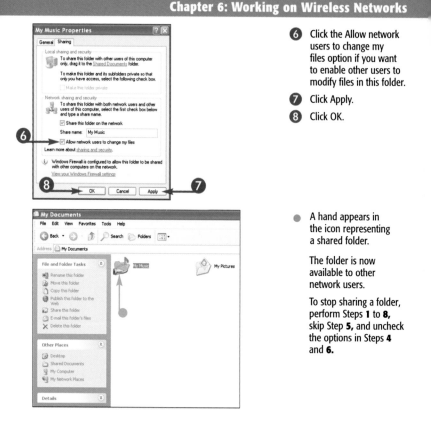

**⑥** Click the Allow network users to change my files option if you want to enable other users to modify files in this folder.

**⑦** Click Apply.

**⑧** Click OK.

● A hand appears in the icon representing a shared folder.

The folder is now available to other network users.

To stop sharing a folder, perform Steps **1** to **8,** skip Step **5,** and uncheck the options in Steps **4** and **6.**

## Did You Know?

You can hide a folder by adding a dollar sign ($) to the end of the share name. For example, you can hide a folder named Photos by renaming it as Photos$. However, network users who know or can guess the folder name can still access the hidden folder. You can assign a drive letter to a hidden folder, making it easier to access on the network.

You can monitor the folders that you share on your wireless network. Windows XP enables you to view which shared folders are being used, the path of the shared folders, and how many users are connected to a shared folder.

You must have administrator privileges to monitor shared folders. As an administrator, you can see a list of shared folders and determine the number of users who are currently accessing the folders.

① Click start.

② Click Control Panel.

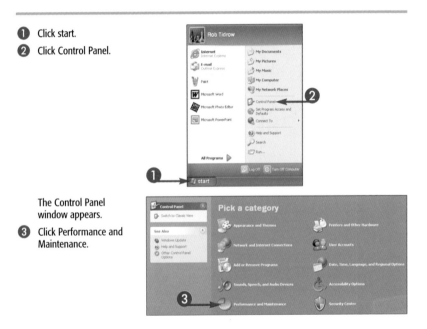

The Control Panel window appears.

③ Click Performance and Maintenance.

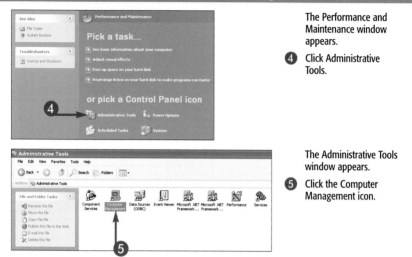

The Performance and Maintenance window appears.

④ Click Administrative Tools.

The Administrative Tools window appears.

⑤ Click the Computer Management icon.

**Did You Know?**

There are some files and folders that you cannot share on your wireless network. You cannot share files in your Documents and Settings, Program Files, and Windows system folders. You also cannot share folders that belong to other users.

Monitoring the number of users who are accessing the folders on your computer helps you, the Administrator, to determine if it is appropriate for you to perform tasks, such as removing or changing shared files, or even turning your computer off.

When others are accessing shared folders and you need to shut down your

computer or do some kind of file maintenance, consider sending e-mails to the users who have access to your shared folders warning the users of the type of work you need to do. It is also best to give the users a specific amount of time — for example, "in 10 minutes" or "at 1:00 P.M." — to allow them to disconnect from the shared folders.

The Computer Management window appears.

**6** Click the plus sign beside the Shared Folders folder.

A list of subfolders appears.

**7** Click Shares.

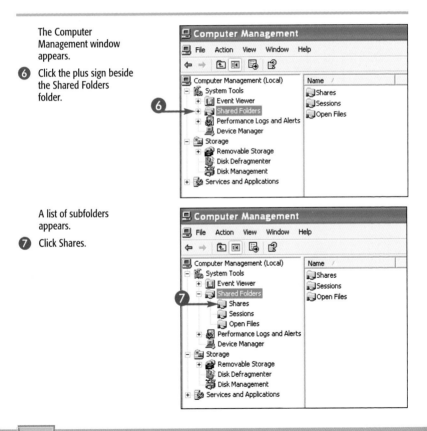

- A list of shared folders appears, listing folder names, folder paths, and the number of users connecting to the shared folders.

- When you finish reviewing the information, you can click the Close icon to close the Computer Management window.

**Did You Know?**

To unshare a folder, navigate to the folder icon, right-click the folder icon, and select Properties. Then click the Sharing tab and select the Share this folder on the network option.

You can assign a drive letter to a folder located on your wireless network. Assigning a letter to a folder makes it easier and faster to access a folder.

Drives letters begin with A and go to Z. Letters A–F are typically used by devices

on your computer: A for a floppy drive, B for a second floppy drive, C for the hard drive, D and E for CD/DVD drives, and F for removable media.

**1** Click start.

**2** Click My Computer.

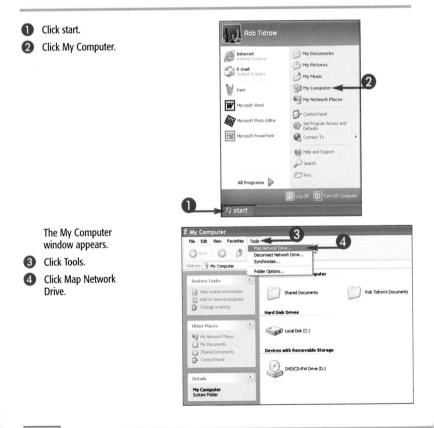

The My Computer window appears.

**3** Click Tools.

**4** Click Map Network Drive.

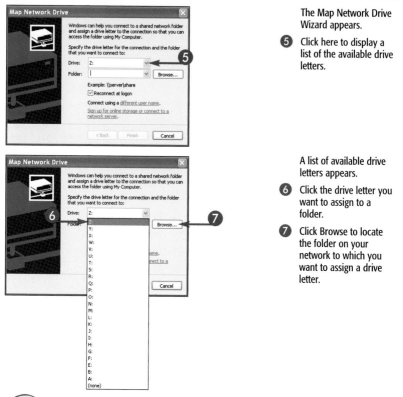

The Map Network Drive Wizard appears.

⑤ Click here to display a list of the available drive letters.

A list of available drive letters appears.

⑥ Click the drive letter you want to assign to a folder.

⑦ Click Browse to locate the folder on your network to which you want to assign a drive letter.

**Did You Know?**

You should use a drive letter from M to Z. You should reserve letters A to J for your floppy disk drive, local hard drives, CD-ROM drives, DVD drives, and other storage devices.

continued

Although drives begin with A and go to Z, typically, network administrators like to start assigning drive letters in reverse order to leave enough contiguous letters for these local devices.

You can access a network folder the same way you access a folder located on your computer. You can easily access a folder shared by computers in your workgroup. A workgroup is a group of related computers on your network.

The Browse For Folder dialog box appears.

Windows automatically highlights your current network workgroup.

⑧ Click the workgroup that contains the folder to which you want to assign a drive letter.

The computers in the workgroup you selected appear.

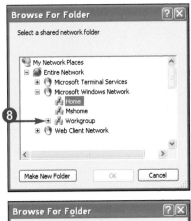

⑨ Click the name of the computer to which you want to assign a drive letter.

The folders on the computer you selected appear.

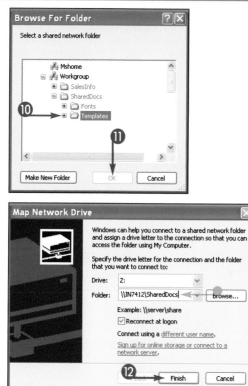

⑩ Click the name of the folder to which you want to assign a drive letter.

⑪ Click OK.

● The Map Network Drive wizard displays the folder you selected.

⑫ Click Finish.

Windows assigns the drive letter to the folder.

You can now access the folder the same way you access local drives, such as a hard drive.

**TIP**

**Did You Know?**

Folders to which you assign a drive letter automatically connect every time you log on to your computer. However, you can configure Windows not to do this if you do not want your computer to automatically connect to the folders. Before performing Step **12** in this section, you can uncheck the box beside Reconnect at logon if you want to disable the feature.

You can use a printer that is attached to another computer on your wireless network. Sharing a printer with other users on your wireless network saves money and space.

To share a printer, the printer must be one that is manufactured to allow sharing.

Although most printers are of this kind, do not assume all of them are. Contact the manufacturer or research the printer before purchasing it to make sure it is one you can share on a network.

① Click start.

② Click Printers and Faxes.

The Printers and Faxes window appears.

③ Click the Add a printer link.

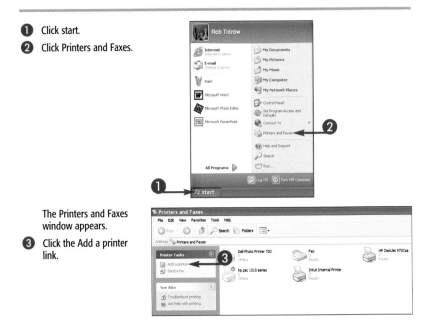

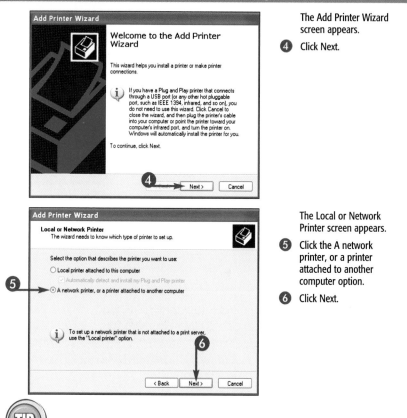

The Add Printer Wizard screen appears.

④ Click Next.

The Local or Network Printer screen appears.

⑤ Click the A network printer, or a printer attached to another computer option.

⑥ Click Next.

**Did You Know?**

You can add and select additional printers, making them available for your use on a wireless network. The first printer you add becomes your default printer, which is automatically used unless you specifically select another printer. Your default printer displays a check mark in its icon.

continued

All the network printers that you add are available for your use. You can use most applications on your computer to print to a network printer. Before you do so, take the time to know exactly where that printer is. That way you do not print documents to a printer that is located in a different building, across campus, or in a

locked office. Also, make sure the printer uses the type of paper on which you need to print. Some printers use plain paper, while others use specialized paper.

If you are using a printer attached to another user's computer, you should get permission before using the printer.

The Specify a Printer screen appears.

⑦ Click the Browse for a printer option.

⑧ Click Next.

⑨ Click an option to add a printer.

For the Browse for a printer option, you can select a printer to add.

For the other two options, you must fill in either the name or the URL for the printer.

⑩ Click Next.

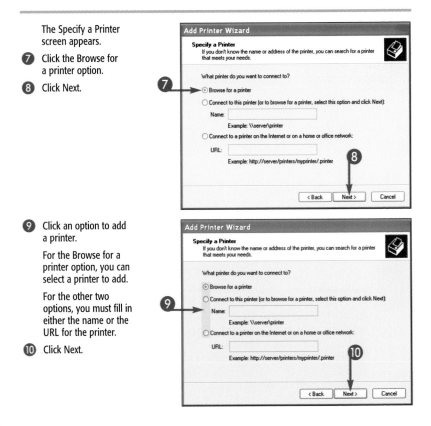

A Windows message box appears, asking you to confirm that you want to add a printer.

⑪ Click Yes.

The Completing the Add Printer Wizard screen appears.

⑫ Click Finish.

Windows adds the printer you selected.

**TIP**

**Did You Now?**
As long as a shared printer is functioning and correctly connected to the wireless network, you can use the printer. Some printers may require intervention, such as adding paper or ink, which prevents the printer from being used until serviced.

# Administering Wireless Networks

Maintenance is an important part of any network. You should perform network maintenance tasks on a daily, weekly, monthly, and annual basis. You may want to create a journal or task log that keeps track of all the administration tasks you must do if you are in charge of a host PC on a wireless network.

Some of the tasks include activating the guest account, changing user passwords, and setting up shares. You also can administer user accounts on Windows XP computers that connect to your wireless network. In this chapter, you learn how to add and delete users, change user passwords, and perform other administrative functions.

# Quick Tips

# Establish User Accounts

Wireless networks typically require all users to have their own user account before they can access the network. User accounts provide the initial way for a network system to guard against unauthorized entry into the network resources.

When you set up accounts on your network, consider how large the user base may become. If a small group of people uses the system, first names or even nicknames may suffice, as long as others who are added later do not share names with existing members. A good system for networks that expect to grow is to use first initial and last name, first name_last name, or a similar naming convention.

## Authentication

The primary purpose of user accounts is to allow the computer to verify that the person accessing the computer is the person who is authorized to do so. The most common way of accomplishing this is to require each user to have his or her own individual login ID, or username, accompanied by a password.

## Restrictions

You can use user accounts to restrict the network services and resources that a user can access. These services and resources may be on the computer on which the user is logged on. Most networks place restrictions on the services and resources that users can access to better manage the network and increase security.

### Special Accounts

There are a number of user accounts on a computer that may not represent an actual person. For example, a computer may have an account called *backup* that is used solely by a backup application. Two other common special accounts are the guest account, for temporary users of a computer, and the administrator account, which manages the computer.

### Groups

On networks, user accounts are typically arranged into groups to make the management of users easier. For example, a group called *Internet* can contain the names of users who may access the Internet. Users not listed in this group cannot access the Internet using the network.

### Administration

User accounts, whether on a computer or a network, require an administrator to create and manage those accounts. An administrator has a special designated user account that gives him or her the capability to manipulate user accounts. On a large network, you can have many administrators managing the network and the user accounts.

# Add a User Account

You can add a user account to a computer to enable different people to access the computer and the wireless network.

When you set up a user account, Windows XP associates settings for that user on the computer. For example, each user can have a unique desktop, start menu, wallpaper, and Internet favorites folder. Each user also can have a different picture for his or her user account.

Each user that wants to log in to your network needs his or her own username and password. You can set up user accounts using the User Accounts program in Windows XP.

1. Click start.
2. Click Control Panel.

The Control Panel window appears.

3. Click User Accounts.

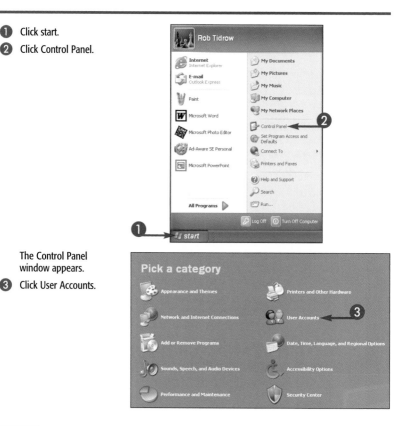

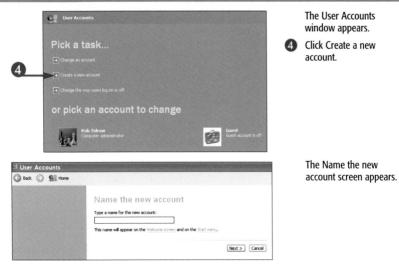

The User Accounts window appears.

④ Click Create a new account.

The Name the new account screen appears.

#### Did You Know?

Not everyone can create a new user account in Windows XP. To create a user account, you must log in using a computer administrator account. You use computer administrator accounts to perform maintenance tasks on the computer, such as managing user accounts.

continued

**135**

Make sure each user on a computer has his or her own user account to help keep files and information, such as e-mail messages, organized.

Windows XP makes it easy to set up multiple users on the same computer. For shared computers like this, set up a different user name for each person. This way, when each user logs in, specific settings for that user, including network settings, and folders, are accessible to that user.

Many companies and schools reduce expenses by using shared computers. Instead of purchasing one computer for each person, an organization can purchase one computer for many people.

---

**5** Type a name for the new account.

**6** Click Next.

> ### Name the new account
>
> Type a name for the new account:
>
> | Tamika |
>
> ← **5**
>
> This name will appear on the Welcome screen and on the Start menu.
>
> **6** → [ Next > ] [ Cancel ]

The Pick an account type screen appears.

**7** Click the type of account you want to create.

***Note:*** *You should create a Limited account type for users who are not computer administrators.*

> ### Pick an account type
>
> ◉ Computer administrator   ○ Limited ← **7**
>
> With a computer administrator account, you can:
> - Create, change, and delete accounts
> - Make system-wide changes
> - Install programs and access all files
>
> [ < Back ] [ Create Account ] [ Cancel ]

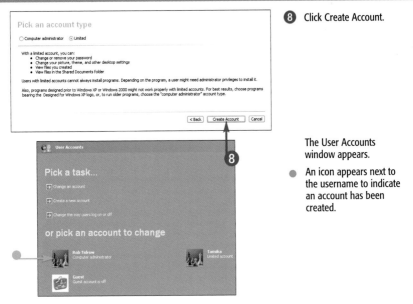

8 Click Create Account.

The User Accounts window appears.

● An icon appears next to the username to indicate an account has been created.

**TIP**

**Did You Know?**
There is no limit to the amount of user accounts you can create. However, you should keep the number of accounts at a manageable level.

# Delete a User Account

Windows XP enables you to create as many users as you want on a computer. However, you should keep the number to a manageable one; say, no more than a half dozen. You can remove a user when you get too many users or have a user who no longer uses this computer.

You can delete accounts for users who no longer need access to your computer or your wireless network to conserve computer resources and increase network security.

① Click start.

② Click Control Panel.

The Control Panel window appears.

③ Click User Accounts.

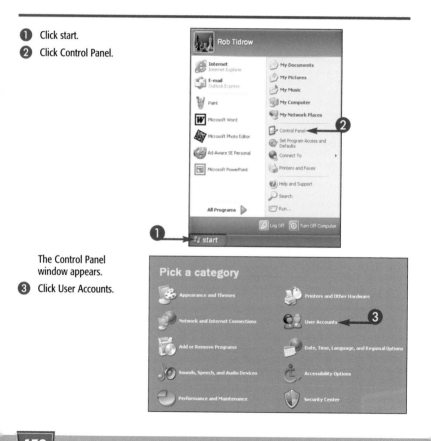

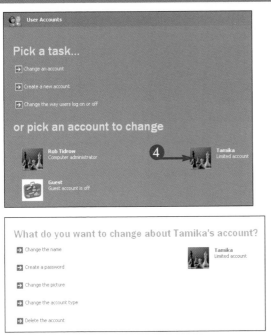

The User Accounts window appears.

④ Click the account you want to delete.

A screen appears asking what you want to change about the selected account.

**Did You Know?**

When you delete a user account, Windows XP asks if you want to delete the files belonging to that user account. If you delete those files, the files belonging to the deleted user account are permanently deleted.

continued

A computer must always have at least one administrator account authorized to create and delete user accounts.

If you work in an area in which a person is responsible for the maintenance and management of computers, have just one administrator for each computer.

However, if the computer is in a remote area where one administrator is not always around to fix problems on a computer, set up several administrators on each computer. This way, more than one person is responsible for the computer maintenance and management.

**5** Click Delete the account.

A screen appears asking if you want to keep or delete the user's personal files.

**6** Click Delete Files.

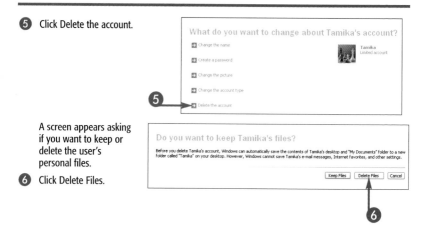

What do you want to change about Tamika's account?

- Change the name
- Create a password
- Change the picture
- Change the account type
- Delete the account

Tamika
Limited account

Do you want to keep Tamika's files?

Before you delete Tamika's account, Windows can automatically save the contents of Tamika's desktop and "My Documents" folder to a new folder called "Tamika" on your desktop. However, Windows cannot save Tamika's e-mail messages, Internet favorites, and other settings.

[ Keep Files ] [ Delete Files ] [ Cancel ]

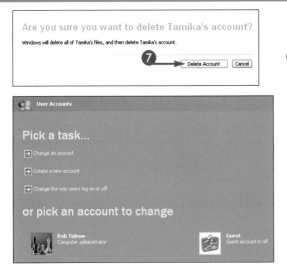

Windows asks you to confirm that you want to delete the account.

**7** Click Delete Account.

The User Accounts window appears displaying the list of remaining user accounts.

The account you removed is gone.

**Did You Know?**

If you choose not to delete the files belonging to a user account that you are deleting, the files belonging to the deleted user account are placed into a folder on the Windows desktop.

# Assign a User Password

You can assign a password for each user account. This helps make your wireless network more secure by preventing unauthorized access to computers connected to the network.

When you choose a password, avoid using passwords that others can easily guess. The most secure passwords have seven or more characters, including lowercase and

uppercase letters, numerals, and other symbols. Avoid using passwords that contain your name or username, or common words or names.

Another habit you should get into is to change your password periodically. Some companies require their employees to change passwords every 30 days, for example.

---

① Click start.

② Click Control Panel.

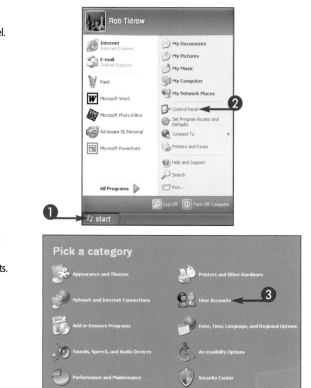

The Control Panel window appears.

③ Click User Accounts.

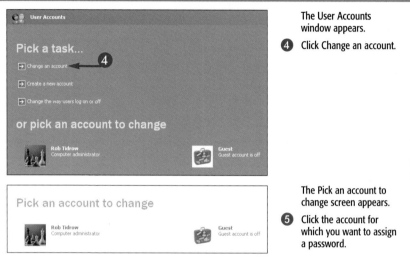

The User Accounts window appears.

④ Click Change an account.

The Pick an account to change screen appears.

⑤ Click the account for which you want to assign a password.

**Did You Know?**
Windows XP does not allow you to access your computer without a password after a user account and password are set up. Unlike Windows versions in the past, such as Windows 98, you cannot click Cancel to bypass the login screen. You must log in with a valid username or password.

continued

143

Try to avoid writing your password down or letting other users know your password. If you must keep a list of your passwords, ensure they are in a safe location.

One thing you may do is write the password and place it in your wallet or purse. Then if you need a reminder of your password, look in these places for it. Do not write it down and place it under your keyboard, in your top desk drawer, or hang it on your monitor. These are the first places people look for it.

A screen appears asking what you want to change about a specific account.

⑥ Click Create a password.

What do you want to change about your account?

➜ Change my name

⑥ ➜ Create a password

➜ Change my picture

➜ Change my account type

➜ Set up my account to use a .NET Passport

**Rob Tidrow**
Computer administrator

The Create a password for your account screen appears.

⑦ Type a new password.

⑧ Type the password again to confirm.

⑨ Type a word or phrase that will help you remember the password.

⑩ Click Create Password.

Create a password for your account

Type a new password:
⑦ ••••••••

Type the new password again to confirm:
⑧ ••••••••

If your password contains capital letters, be sure to type them the same way every time you log on.

Type a word or phrase to use as a password hint:
⑨ The  town where you were born

The password hint will be visible to everyone who uses this computer.

⑩ ➜ Create Password   Cancel

Do you want to make your files and folders private?

Even with a password on your account, other people using this computer can still see your documents. To prevent this, Windows can make your files and folders private. This will prevent users with limited accounts from gaining access to your files and folders.

Yes, Make Private    No

What do you want to change about your account?

→ Change my name

→ Change my password

→ Remove my password

Rob Tidrow
Computer administrator
Password protected

→ Change my picture

→ Change my account type

→ Set up my account to use a .NET Passport

A screen appears asking if you want to make your files and folders private.

⑪ Click Yes, Make Private.

This protects your files from other users of the computer.

A screen appears asking what you want to change about a specific account.

**Did You Know?**

If you cannot remember your password and therefore cannot log in to your account, a user with a computer administrator account will have to delete the password from your user account. After it is deleted, you can create a new password.

The Guest account is a permanent account that is added to your computer when you install Windows XP. You cannot remove it, but you can inactivate it. In fact, when you first install Windows, the Guest account is inactive.

You may find times when it is good to have it turned on. You can activate the

Guest account on your Windows XP computer to enable temporary users to access your computer and network. Another reason to activate the Guest account is to use it to test your network's file and printer sharing settings after you get those items set up.

1 Click start.

2 Click Control Panel.

The Control Panel window appears.

3 Click User Accounts.

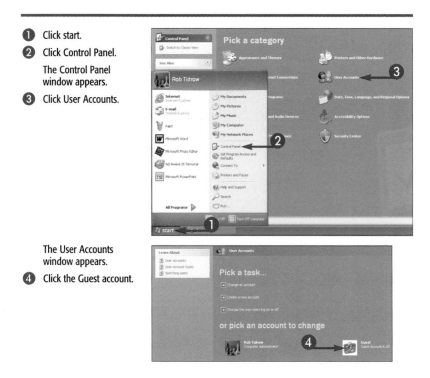

The User Accounts window appears.

4 Click the Guest account.

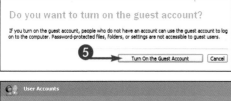

A screen appears asking if you want to turn on the Guest account.

⑤ Click Turn On the Guest Account.

● A screen appears indicating the Guest account is now active.

**Did You Know?**

Users who log on to your computer using the guest account cannot access password-protected files, folders, and settings. For security reasons, you may want to turn on the guest account only when necessary.

# Change a Workgroup Name

Wireless networks must have workgroups to enable users to connect to the resources available on it. Workgroups are analogous to departments — such as administration, sales, and marketing — in a large company.

The workgroup name must be the same on all computers on a wireless network. If necessary, you can change the workgroup name on a computer that you add to a wireless network so it can easily communicate with the other computers on the network.

If you are connecting to a wireless network and are not sure of the workgroup name, ask someone what it is. Without the workgroup name, you cannot access shared resources on the network.

① Click start.

② Click Control Panel.

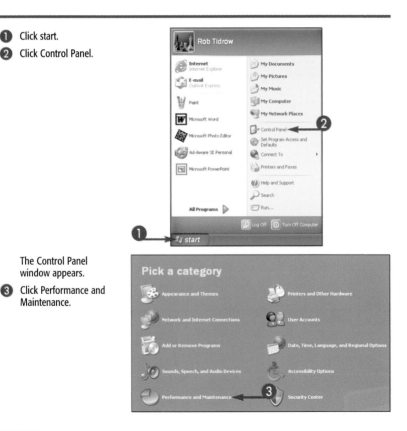

The Control Panel window appears.

③ Click Performance and Maintenance.

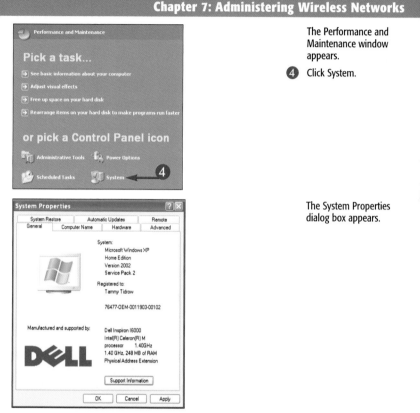

The Performance and Maintenance window appears.

④ Click System.

The System Properties dialog box appears.

**TIP**

**Did You Know?**
When choosing a workgroup name, try to keep workgroup names logical. For example, try using *sales* for the sales department or *home* for a home network. Make sure your workgroup names are short and do not contain spaces.

continued

Wireless networks can consist of computers using multiple workgroup names. You can change the workgroup name of your computer to access the resources of the different workgroups.

If you travel a great deal with a laptop and connect to wireless networks in various places, you need to become comfortable

changing workgroup names. Each network will have different users attached to it and you probably will want to connect to those resources. To access those resources, you must know which resource uses what workgroup name and then change your settings to match that name.

**⑤** Click the Computer Name tab.

**⑥** Click Change.

The Computer Name Changes dialog box appears.

**⑦** Click in the Workgroup field and edit the workgroup name.

**⑧** Click OK.

Windows accepts the new workgroup name.

A dialog box appears, welcoming you to your new workgroup.

**9** Click OK.

A dialog box appears informing you that you must restart the computer.

**10** Click OK.

The workgroup name is changed.

*Note: You must reboot your computer for the changes to take effect.*

**Did You Know?**

Larger computer networks use domains instead of workgroups to organize computers. If your computer requires you to join a domain, you must consult with the network administrator for instructions on how to access the domain.

# Share Information on a Wireless Network

After you connect to a wireless network, you can share information in your folders on your computer with other users on the wireless network.

You must set up the folders you want to share by specifying each one. If you want to share several folders, but do not want

to set up shares for each one individually, create one primary folder and then move the other folders into this primary one. This makes the moved folders subfolders. When you share the primary folder, all subfolders are shared as well.

① Click start.
② Click My Documents.

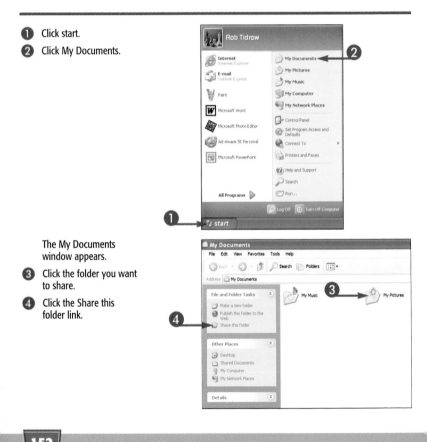

The My Documents window appears.
③ Click the folder you want to share.
④ Click the Share this folder link.

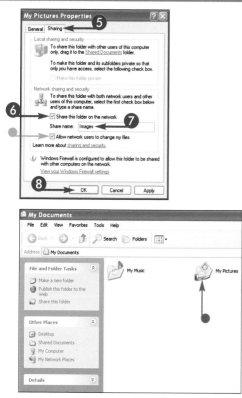

The Properties dialog box for the folder you selected appears.

⑤ Click the Sharing tab.

⑥ Click the Share this folder on the network option.

⑦ Type the name of the shared folder as it will appear on the network.

● You can click the Allow network users to change my files option if you want to allow others to change the files.

⑧ Click OK.

● A hand appears under the folder icon to indicate a shared folder on the wireless network.

**TIP**

**Did You Know?**

Allowing users to change your files lets them delete files as well. It also makes your files vulnerable should someone gain unauthorized access to your wireless network. Only allow users to change your files if the information is not essential or if you have current backups of the shared information.

153

One way to save money in your organization or home is to use one printer among several users. You can share your printer that is attached to your computer with other users on a wireless network.

When you share a printer on a wireless network, you attach the printer to one computer. You then set up the printer as a shared printer so others can connect to it over the network. The other computers must configure Windows to work with this shared printer.

A shared printer shows up in your Printers and Faxes folder with a hand icon next to the printer icon.

① Click start.

② Click Printers and Faxes.

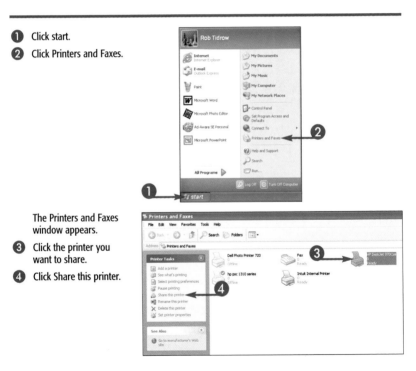

The Printers and Faxes window appears.

③ Click the printer you want to share.

④ Click Share this printer.

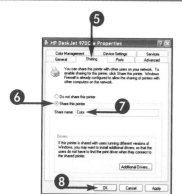

The Printer Properties dialog box appears.

⑤ Click the Sharing tab.

⑥ Click the Share this printer option.

⑦ Type the name of the printer as you want it to appear on the network.

⑧ Click OK.

● A hand appears under the printer icon to indicate a shared printer on the wireless network.

**TIP**

**Did You Know?**

Most operating systems have an easy way for users to locate and print to network printers—even those on a wireless network. To access a printer using a Windows XP computer, users can use the Add a Printer Wizard available from the Printers and Faxes window task pane.

# Securing Wireless Networks

One of the greatest concerns with wireless networks is how to keep data and resources secure. By doing a few simple things, you can make your wireless network more secure. This chapter introduces such security features as WEP encryption, SSID names, and firewalls. These features, along with firewalls and technologies you learn about as you become more wireless savvy, can help keep out unwanted and unauthorized users.

Your wireless gateway probably has a built-in firewall, but it only stops malicious Internet traffic from entering your network. With other software, such as Windows Firewall, you can stop unauthorized traffic originating from both the Internet and from within your own network.

# Understanding Network Security

When you set up and use a network, including a wireless network, consider creating a security policy. This policy is to be enforced on all users who use your network. There are many components to a good security policy. All network administrators should evaluate each aspect of their network security.

Some of the policies you may consider include data encryption, unique hardware identification, Internet restrictions, and data filtering. Almost all medium to large companies are required (usually through shareholders or corporate officers) to document and adhere to strict security policies.

## Encryption

A common way to protect information during its transmission, regardless of how the information is transmitted, is to encrypt it. *Encryption* takes the information and scrambles it so that it is unreadable. Only the intended recipient of the information has the code that allows the data to be unscrambled.

## Identification

Each piece of equipment connected to a network is unique and is identifiable by its unique identification number required to connect to the network. It is possible to restrict network access to only those devices of which you are aware. This prevents any unauthorized equipment, such as another computer, from accessing your network without specific permission.

## Filter Data

All data on a network is formatted into very specific types of data. For example, the information in an e-mail message is very different from the information you use to log on to the network. You can restrict the type of data on a network to prevent unauthorized users from using different data types to access the network. For example, if you never access the World Wide Web on your network, you can disable the transmission of Web pages over the network. Filtering programs also enable you to filter the type of content users can access over the Internet. For example, you can restrict access to Web sites that meet certain criteria, such as violence, sexual themes, or drugs. Filtering software also lets you track user activity, including which user visited a specific site.

## Restrict Sources

All information on a network contains the identification of the computer sending the data and the identification of the computer receiving the data. If needed, you can restrict the information that transfers on your network depending on the identification of the source. For example, you can restrict access to files located on the Internet. This prevents users on the Internet from attempting to access your network files. You also can restrict which computers can transfer files to your network, reducing potential threats to your network's security.

Implementing a sensible security policy on your wireless network has two major benefits: secure resources and protected data.

When you create a security policy, be ready to present it to all your employees or users. In addition, have regular meetings to discuss, update, and amend security policies as your network grows and becomes more popular. With an up-to-date security policy, companies can limit the amount of liability they may be responsible for, especially if they store sensitive client or customer data.

You should also audit the security policy over time to ensure it is still effective.

### Secure Resources

One of the primary reasons that people make attempts to access wireless networks is so that they can obtain access to the Internet. Without a proper security policy in place, unauthorized users can use your wireless network, thereby restricting the speed at which your network can transmit information.

### Data Protection

Unauthorized access to your network leaves your data open to others. Apart from actually deleting or otherwise corrupting your data, the unauthorized user may copy your sensitive data and use it inappropriately, such as selling credit card numbers you stored on your computer to others.

There are two primary disadvantages to implementing a comprehensive security policy: management and usage speed.

As with any type of rule or set of rules, security policies can get in the way of some users who want to access and use your network. A user may not want to jump through all the hoops just to use your network. In fact, some users may not use your network over time if the security policy is too rigid.

Do not be discouraged by this, especially if you do not want your network resources and data comprised.

## Management

Planning, implementing, and maintaining a security policy is a very time-consuming process requiring the attention of the network administrator. Not only do you have to implement the security policy, but you must also stay abreast of the latest developments in wireless security to ensure your network does not become vulnerable.

## Speed

One of the trade-offs of implementing a security policy is the speed at which the network operates and a user can work. If it is necessary to encrypt data, it takes longer for the data to travel between computers. Requiring users to use a password to access resources such as the computer files and printers reduces the speed at which users can work.

One of the benefits, and at the same time a disadvantage, of wireless networks is that they broadcast information over a large geographic area, sometimes in excess of two or three miles. While this enables devices to communicate easily in many different locations, it also makes it easier for unauthorized persons to attempt to access that information. Most persons trying to access your wireless network are looking for access to the Internet, but some may want to access your network's data.

Because wireless networks work off of radio signals, anyone in your area with a receiving device (a laptop computer with a wireless card, for instance), can receive those signals.

## Lax Security Procedures

Wireless networks all use some security in the form of passwords and identifications that allow only specific, authorized devices to connect to the wireless network. Unfortunately, many network administrators fail to take the time to properly implement a wireless network security process, making it that much easier for security breaches to happen.

## Eavesdropping

By far the most common threat to a wireless network is in the form of eavesdropping on the signal, which the wireless network generates and which extends beyond the property of the network owner. For example, a wireless network in an office may transmit signals through the building wall to a location next door. The equipment an intruder can use to eavesdrop on wireless networks is the same that you use to construct the network and is available at most computer stores.

### Jamming

Another threat, although not as common as eavesdropping, is jamming of a wireless network. Jamming involves using a transmitter to broadcast signals to a device or devices of a wireless network from a nearby location in an attempt to overload the wireless network and cause it to fail. Again, the equipment used to jam a wireless network is readily available. If you think you are experiencing jamming of your network, consult a networking or security expert, usually called a networking security specialist. They can analyze your network to determine the threat and offer solutions.

### Reducing Threats

Nothing reduces the threats to a wireless network better than carefully planning your wireless network installation. Consult the documentation that comes with your wireless network equipment and fully implement any security measures that the hardware manufacturer recommends. If you network has particularly sensitive data on it, then it is best to consult with a network security specialist before implementing a wireless network. Networking security specialists can perform tests and analysis on your network to determine the types of threats your system is vulnerable to. In fact, these experts can show you first hand, usually through a live demo on your system, how they can access information that you may have initially thought was tightly secured.

You can use Wired Equivalent Privacy (WEP) encryption to scramble the data transferred using a wireless network. This prevents unauthorized users from viewing information as it moves across your wireless network.

The example in this section illustrates how to use a Web browser to configure

a wireless device from Microsoft. Remember to follow your wireless device instructions on opening the configuration settings for your device.

For more information on setting up a wireless gateway and accessing the Base Station Management Tool screen, see Chapter 2.

① In the Base Station Management Tool screen, click Security.

② Click Wireless Security.

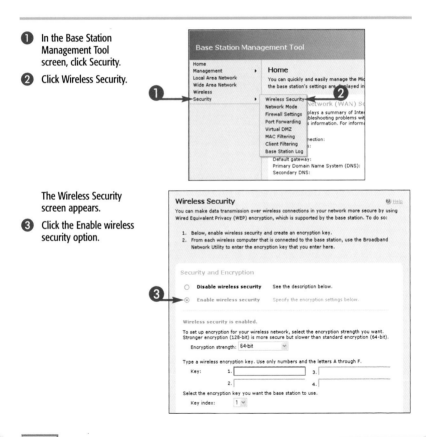

The Wireless Security screen appears.

③ Click the Enable wireless security option.

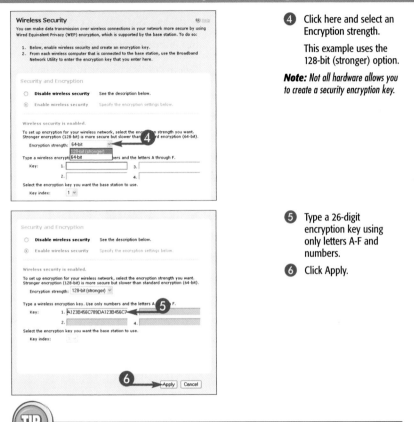

**Wireless Security**

You can make data transmission over wireless connections in your network more secure by using Wired Equivalent Privacy (WEP) encryption, which is supported by the base station. To do so:

1. Below, enable wireless security and create an encryption key.
2. From each wireless computer that is connected to the base station, use the Broadband Network Utility to enter the encryption key that you enter here.

Security and Encryption

○ **Disable wireless security**   See the description below.

◉ **Enable wireless security**   Specify the encryption settings below.

Wireless security is enabled.

To set up encryption for your wireless network, select the encryption strength you want. Stronger encryption (128-bit) is more secure but slower than standard encryption (64-bit).

Encryption strength:   64-bit

Type a wireless encryption key. Use only numbers and the letters A through F.

Key:   1.                    3.
       2.                    4.

Select the encryption key you want the base station to use.

Key index:   1

---

Security and Encryption

○ **Disable wireless security**   See the description below.

◉ **Enable wireless security**   Specify the encryption settings below.

Wireless security is enabled.

To set up encryption for your wireless network, select the encryption strength you want. Stronger encryption (128-bit) is more secure but slower than standard encryption (64-bit).

Encryption strength:   128-bit (stronger)

Type a wireless encryption key. Use only numbers and the letters A through F.

Key:   1. A123B456C789DA123B456C7
       2.                    4.

Select the encryption key you want the base station to use.

Key index:

Apply   Cancel

**④** Click here and select an Encryption strength.

This example uses the 128-bit (stronger) option.

***Note:*** *Not all hardware allows you to create a security encryption key.*

**⑤** Type a 26-digit encryption key using only letters A-F and numbers.

**⑥** Click Apply.

---

**TIP**

**Did You Know?**

You can choose between 64-bit and 128-bit encryption. Use the highest encryption level your equipment allows. Keep in mind that your network may operate more slowly at the 128-bit setting because of the additional processing power the higher encryption level requires.

continued

You can enable WEP encryption on your wireless network once the network is operating correctly. Make careful note of the WEP settings you choose. You cannot access the wireless network without them.

Most small wireless networks do not use WEP to protect them. This is fine as long as the data and resources that are available

are limited. If you do not plan to use WEP, you should share only those folders that include nonsensitive information in them. Without WEP protection, do not share folders on wireless networks that include personnel files, private company information, trade secrets, and financial data.

⑦ Right-click the Network Connection icon in the system tray.

⑧ Click Status.

The Wireless Network Connection Status dialog box appears.

⑨ Click Properties.

The Wireless Network Connection Properties dialog box appears.

⑩ Click the Wireless Networks tab.

⑪ In the Preferred networks area, click Properties.

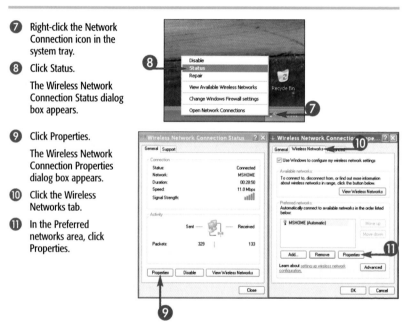

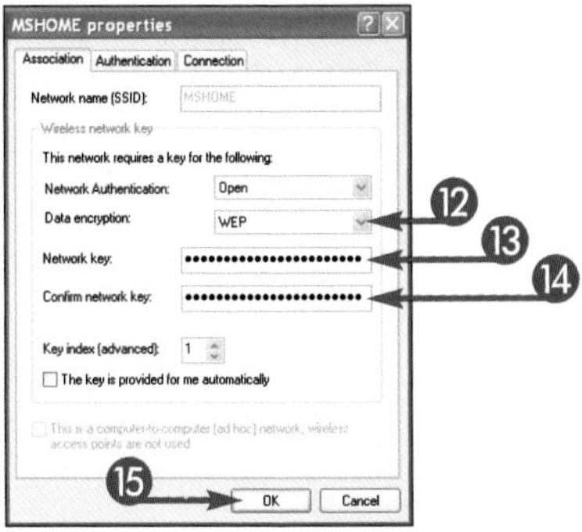

The properties dialog box for your network appears.

⑫ Click here and select WEP from the Data encryption list.

⑬ Type the key text generated in Step **5.**

⑭ Retype the key text.

⑮ Click OK.

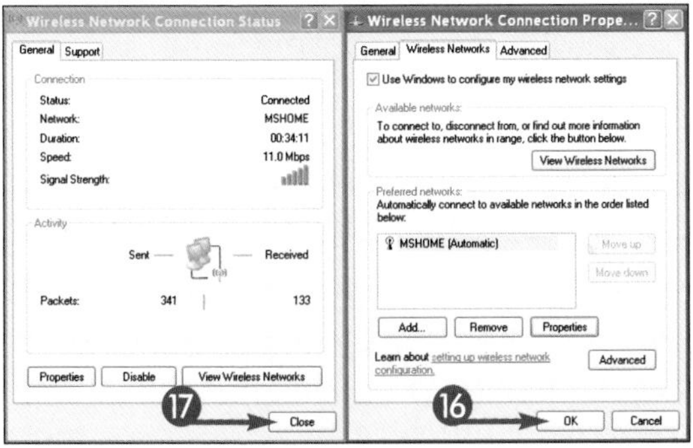

⑯ Click OK to close the Wireless Network Connection Properties dialog box.

⑰ Click Close to close the Wireless Network Connection Status dialog box.

The wireless network connection now uses WEP.

**TIP**

**Did You Know?**

You should use WEP on all wireless networks. Information you may consider harmless, such as street addresses, phone numbers, or even accurate birth dates, is often stored on computers. People who specialize in identity fraud seek this type of apparently harmless information.

The SSID, or *service set identifier*, is the name used to identify your wireless network. When you initially configure a wireless network, you should change the SSID of the network to make it harder for unauthorized persons to access your wireless network.

The steps in this section show how to use a Web browser to configure a wireless router from Microsoft. Remember to follow your wireless device instructions on opening the configuration settings for your device.

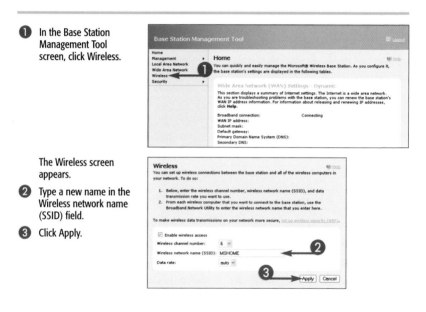

① In the Base Station Management Tool screen, click Wireless.

The Wireless screen appears.

② Type a new name in the Wireless network name (SSID) field.

③ Click Apply.

Windows displays a screen reminding you to configure each computer on your network with the new name.

④ Click OK.

Your router updates its settings.

**Microsoft Internet Explorer**

? Remember, you must also configure each computer that connects wirelessly with the base station.

The base station and each computer that connects to it must use the same wireless network name and wireless channel number.

④ OK    Cancel

**Wireless**

You can set up wireless connections between the base station and all of the wireless computers in your network. To do so:

1. Below, enter the wireless channel number, wireless network name (SSID), and data transmission rate you want to use.
2. From each wireless computer that you want to connect to the base station, use the Broadband Network Utility to enter the wireless network name that you enter here.

To make wireless data transmissions on your network more secure, set up wireless security (WEP).

☑ Enable wireless access

Wireless channel number:  6 ∨

Wireless network name (SSID):  MS_HOME

Data rate:  auto ∨

Apply    Cancel

**Did You Know?**

You may be able to prevent others from viewing your SSID. If your router provides a disabling feature, you can turn off SSID name broadcasting. Many routers provide this feature, but it means that all users connecting to your network must know the name, or SSID, of the wireless network.

# Manage a Hardware Access List

The MAC, or Media Access Control, address is a unique hardware identification number that every wireless device that connects to a wireless network must have. You can create a Hardware Access List that contains the MAC addresses of the network hardware that can connect to your wireless access point.

The steps in this section illustrate how to use a Web browser to configure a wireless router from Microsoft. Remember to follow your wireless device instructions on opening the configuration settings for your device.

① In the Base Station Management Tool screen, click Wireless.

**Note:** To access this screen as well as to learn more about setting up a wireless gateway, see Chapter 2.

② Scroll down to the end of the screen.

③ Click Refresh.

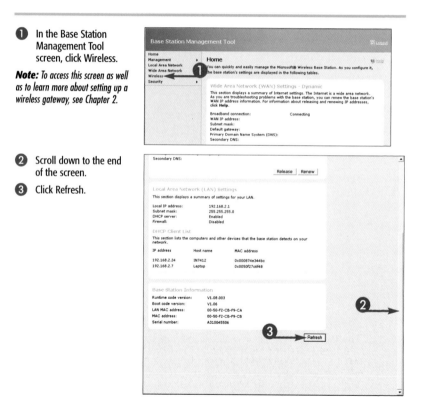

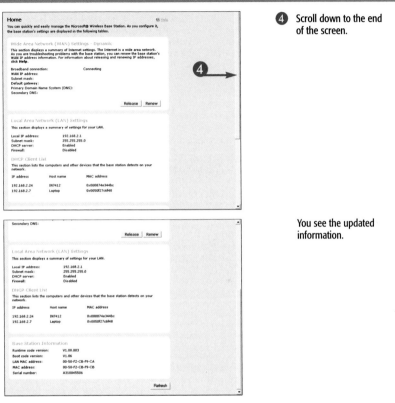

④ Scroll down to the end of the screen.

You see the updated information.

### Did You Know?

If your router software lets you, you can block a specific piece of wireless networking hardware from connecting to your network by deleting its MAC address from the Access Control List. For example, you may want to block a MAC address that belongs to another person who uses the same wireless network as you, such as a neighbor.

# Configure Windows Firewall

You can install firewall software that prevents unauthorized Internet traffic from entering or leaving the computers on your network. A firewall prevents unauthorized users from gaining access to your wireless network. Windows Firewall is one of the most used firewall applications and is available with Windows XP.

The basic way a firewall works is by masking the true identity of the computers on your network. The firewall becomes a "proxy" for your computer to the outside world (the Internet). Users on the Internet can see the firewall address, but cannot see behind it to view individual computers on the network.

① Click start.

② Click Control Panel.

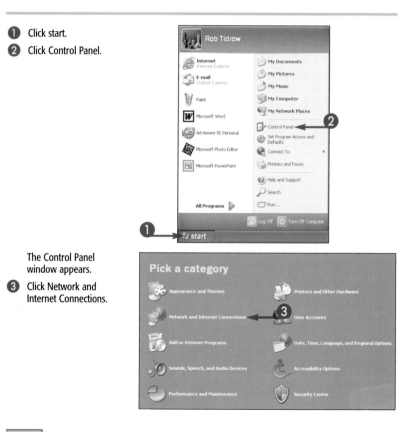

The Control Panel window appears.

③ Click Network and Internet Connections.

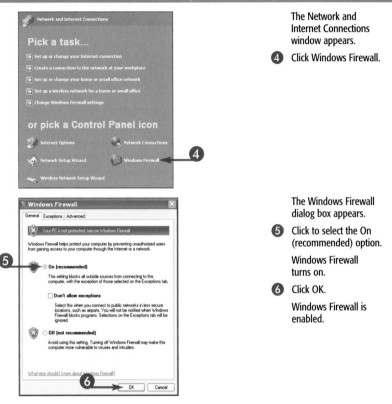

The Network and Internet Connections window appears.

④ Click Windows Firewall.

The Windows Firewall dialog box appears.

⑤ Click to select the On (recommended) option.

Windows Firewall turns on.

⑥ Click OK.

Windows Firewall is enabled.

---

**TIP**

**Did You Know?**

There are many other firewall products available, including some that are free. You can download firewall software from Sunbelt Software, which offers home users a free version of Sunbelt Kerio Personal Firewall at www.sunbelt-software.com/Kerio.cfm.

Windows Firewall blocks users trying to access your computer using Windows Remote Desktop. Your wireless gateway probably has a built-in firewall, but it may be limited in its features. Windows Firewall enables you to set up exceptions,

which lets you turn on the firewall for all but those programs and services you want to allow access to your computer. This is handy when you know the firewall blocks programs and services that you want to use, such as file and printer sharing.

① Start the Windows Firewall application.

*Note: See the section "Configure Windows Firewall" to open the Windows Firewall dialog box.*

② Click to deselect the Don't allow exceptions option.

③ Click the Exceptions tab.

④ Click to select Remote Desktop.

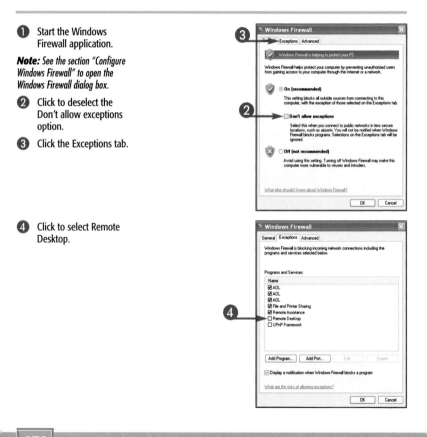

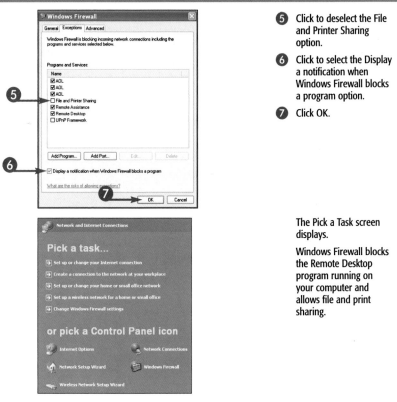

⑤ Click to deselect the File and Printer Sharing option.

⑥ Click to select the Display a notification when Windows Firewall blocks a program option.

⑦ Click OK.

The Pick a Task screen displays.

Windows Firewall blocks the Remote Desktop program running on your computer and allows file and print sharing.

---

**TIP**

**Did You Know?**

Windows Firewall includes a log file tool that helps you troubleshoot problems. Turn it on by opening the Windows Firewall program. Click the Advanced tab and click Settings under the Security Login area. Click Log dropped packets. Click OK, and then click OK again.

# Disable Windows Firewall

Windows Firewall is easy to use and offers a tremendous amount of control over the programs and services that run on your computer across a wireless network. The disadvantage of Windows Firewall is that it can block too much traffic, sometimes rendering your network and Internet connections worthless. You may want to disable Windows Firewall to allow you to work on the network.

A good time to disable Windows Firewall is after you install a network or Internet program and the program does not operate correctly. Temporarily disable Windows Firewall and test the program again. If it works fine, Windows Firewall may need to be set to allow that program to operate.

① Click start.

② Click Control Panel.

The Control Panel window appears.

③ Click Network and Internet Connections.

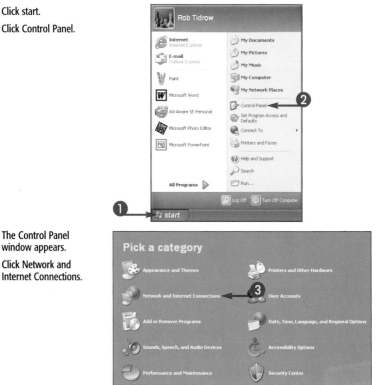

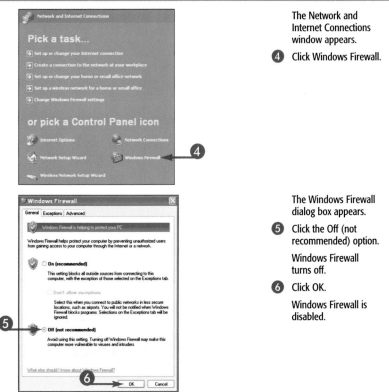

The Network and Internet Connections window appears.

④ Click Windows Firewall.

The Windows Firewall dialog box appears.

⑤ Click the Off (not recommended) option.

Windows Firewall turns off.

⑥ Click OK.

Windows Firewall is disabled.

**Did You Know?**

Windows Firewall is not a cure-all to the problems you may encounter with unauthorized programs and services running on your computer. You must follow good network protection procedures to protect yourself. This includes using a strong password, monitoring users who have access to your computer, and keeping the number of shared folders to a minimum.

# Connecting on the Road

The main reason laptop computers are popular is because of their portability. You can carry them with you to almost any destination. As laptops become more popular, businesses, hotels, schools, and metropolitan areas are providing public access to wireless networks. It is not uncommon to see advertisements for wireless Internet when you shop for hotel reservations.

You can connect your laptop computer to wireless networks when you travel, providing you with Internet access or a direct connection to an office network. You can connect to Wi-Fi networks in airports, coffee shops, and other public areas.

# Quick Tips

Public locations that provide wireless network access allow you to use your mobile computer in places other than work or at home.

As more and more users and workers rely on mobile technologies, wireless networks will grow and become more prevalent. In some metropolitan areas, for example, wireless networks are available to any person who has a computer and a wireless network card.

Large hotel chains, shopping malls, and many restaurants are providing access to wireless networking resources. For example, some restaurant chains sell networking time just like they do a cup of coffee.

## Internet Access

The primary purpose of connecting to a wireless network in a public location is to gain Internet access. Once a mobile device has Internet access, the user is free to use the mobile device for applications that allow people to communicate, such as sending and receiving e-mail messages.

## Socializing

You can also use wireless networks for nonbusiness reasons, such as using the wireless network to meet and communicate with others. Just like chat rooms and other communications forums on the Internet, you can easily communicate with other users on the wireless network to which you are attached.

### Network Connections

Public wireless networks also allow a user to form a connection with another computer connected to the Internet using a secure connection called a virtual private network (VPN). A VPN allows users to access information stored on computers and networks located at their home or, more commonly, at work.

### Hotspots

A *hotspot* is the name given to a location that has wireless network access. The hotspot may be limited to a single room, such as the lounge in a hotel, or it can be much larger, such as within city limits.

### Service Providers

Many companies provide wireless access. Wireless service providers install the necessary hardware that provides coverage to the location such as a café or airport. Most installations are permanent, but some are temporary, such as wireless networks provided for clients of trade shows or conferences. More than one provider may serve a single location.

Many companies are now providing wireless network access to many locations around the world. Wireless access is available in more places due to better standards, cheaper hardware, faster connection times, and lower access fees than in the past.

When you want to connect to a service provider, you usually must install software or modify your existing wireless network settings. This ensures your computer is set up with the same protocols, a network name, and a proper username to access the network.

Before leaving on business trips, many companies now require their workers to learn about wireless networks in the area they plan to visit. This provides them with a way to communicate with the office while away.

## Standards

Wireless services most commonly use a wireless standard referred to as 802.11b, which provides network speeds of 11 Mbps. Some locations will also use the 802.11g standard, which has speeds of up to 54 Mbps. Most wireless network adapters work with the wireless standards used at hotspots.

## Cost

Most wireless networks require you to pay a fee to access the network. Some service providers require you to pay a fee up front, while other locations may allow you to use the wireless network to purchase or use a product purchased from the location where you use the service.

### Configuration

Some service providers may require that you use their custom connection software to connect to their wireless networks. Most wireless service providers allow you to connect to their wireless network by entering the correct name of the wireless network into the configuration of your wireless network adapter on your mobile computer.

### Hardware

Most service providers allow you to connect with a laptop, which has a wireless network adapter, as well as with a handheld computer. Typically, you have restrictions as to the operating system you use on the mobile device. Some networks may not allow access using handheld computers.

### Roaming

*Roaming* is the ability to sign up as a client of one service provider and use the services of another service provider, which gives the user more wireless network hotspots without having to sign up with many different providers. Unlike mobile telephone providers, wireless access service providers are not capable of providing comprehensive roaming services, but may provide them in the future.

A *hotspot* is a public wireless network access point. Although more and more are becoming available, not all hotspots use the same settings and names.

You can find wireless networks using a hotspot directory. For example, the Wi-Fi Alliance, an industry trade group, maintains the Wi-Fi Zone Finder Web

site, a directory of wireless network providers and their locations.

Before you head to a new city, use the Wi-Fi Zone Finder Web site to locate Wi-Fi zones. This way you can map out where you are most likely to connect to a wireless network when you arrive at your destination.

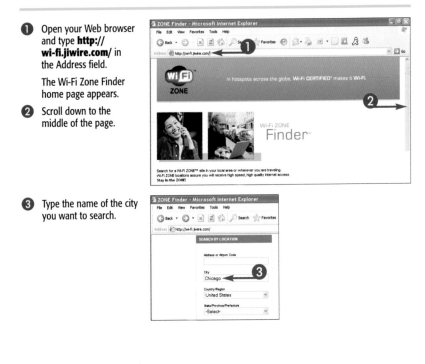

① Open your Web browser and type **http://wi-fi.jiwire.com/** in the Address field.

The Wi-Fi Zone Finder home page appears.

② Scroll down to the middle of the page.

③ Type the name of the city you want to search.

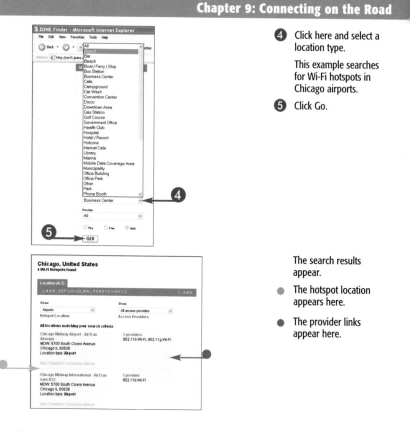

④ Click here and select a location type.

This example searches for Wi-Fi hotspots in Chicago airports.

⑤ Click Go.

The search results appear.

● The hotspot location appears here.

● The provider links appear here.

**Chicago, United States**
4 Wi-Fi Hotspots found

Location (A-Z)

0-9 # A B , D E F G H I J K L M N O , P Q R S T U V W X Y Z          1 - 4 of 4

Show                                    Show
Airports                                All access providers
Hotspot Location                        Access Providers

All locations matching your search criteria

Chicago Midway Airport - Air Tran          2 providers
Airways                                    802.11b Wi-Fi, 802.11g Wi-Fi
MDW: 5700 South Cicero Avenue
Chicago IL 60638
Location type: Airport

Map | Directions | Connection Options

Chicago Midway International - Air Tran     1 provider
Gate B22                                    802.11b Wi-Fi
MDW: 5700 South Cicero Avenue
Chicago IL 60638
Location type: Airport

Map | Directions | Connection Options

**TIP**

**Did You Know?**

You can use an Internet search engine, such as Google, to find the latest Wi-Fi directory resources. To do this, open your Web browser and go to www.google.com. In the search field, type **Wi-Fi hotspots**. The Google search results page appears with a list of links to Wi-Fi hotspot sites.

When you travel, your laptop computer, or other Wi-Fi-compatible device, identifies wireless networks that are broadcasting their presence. Your computer displays a list of these available networks, and you can connect to one of them.

Keep in mind that hotspots come and go. A spot you linked to last month may not be available this month. Or a newer one may be available that offers richer content and faster connections. This is why a Web site like the Wi-Fi Zone is so handy. It does a great job of keeping track of those newer hotspots that you may not be familiar with yet.

① Right-click the Network Connection icon.

② Click View Available Wireless Networks.

The Wireless Network Connection window appears.

③ Click an available network.

④ Click Connect.

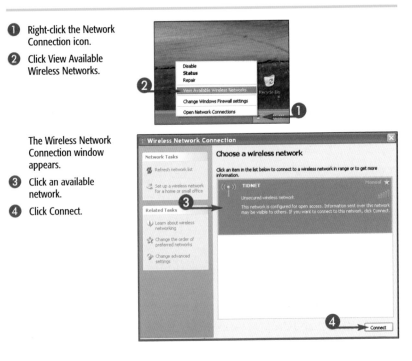

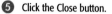

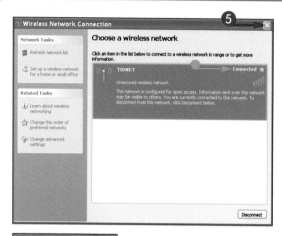

- The status of the network changes to Connected.

**5** Click the Close button.

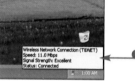

- A pop-up window appears displaying the wireless network connection details.

**TIP**

### Did You Know?

There are national Wi-Fi network access providers you can connect to. Boingo Wireless is one such provider. You can find more information about Boingo Wireless at www.boingo.com. You can also go to a search engine such as Google and type **national Wi-Fi provider** in the search field to display a list of other providers.

You can access your corporate network using a *virtual private network* (VPN) connection. A VPN creates a secure connection between your computer and a VPN server by securely connecting through your wireless network and the Internet.

Many corporations use VPNs to enable computers that are not permanently connected to their network gain access to their networks in a secure way. Although you can use a dial-up modem, VPNs replace standard dial-up connectivity, which may become a security problem for many networks. Companies can use strict security measures to protect VPN connections.

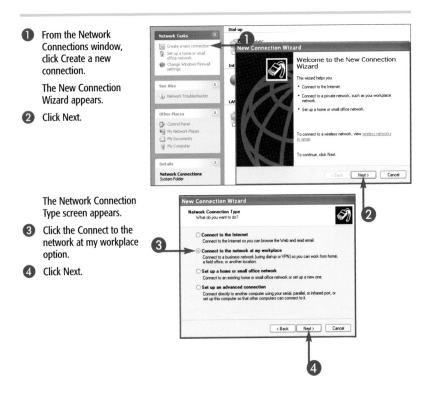

**①** From the Network Connections window, click Create a new connection.

The New Connection Wizard appears.

**②** Click Next.

The Network Connection Type screen appears.

**③** Click the Connect to the network at my workplace option.

**④** Click Next.

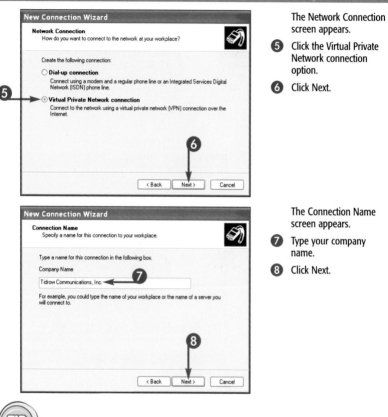

The Network Connection screen appears.

⑤ Click the Virtual Private Network connection option.

⑥ Click Next.

The Connection Name screen appears.

⑦ Type your company name.

⑧ Click Next.

**TIP**

**Did You Know?**

Your computer also can accept incoming VPN connections. You can enable incoming VPN connections with the New Connection Wizard. Other users on your wireless network can then use a VPN connection to connect securely to your computer.

You can connect to your VPN server while traveling, which enables you to access information on your work computer from your mobile computer. One main use of VPN connections is access to corporate-wide e-mail servers. While users are out of the office, such as on business trips or vacation, they can

still connect to the VPN server and access their e-mail and calendar.

Another tool that VPNs take advantage of is the Windows Remote Desktop application. With a VPN connection, users can access their desktop computer via a remote login through the VPN.

The Public Network screen appears.

⑨ Click the Do not dial the initial connection option.

⑩ Click Next.

The VPN Server Selection screen appears.

⑪ Type the name of your VPN host.

⑫ Click Next.

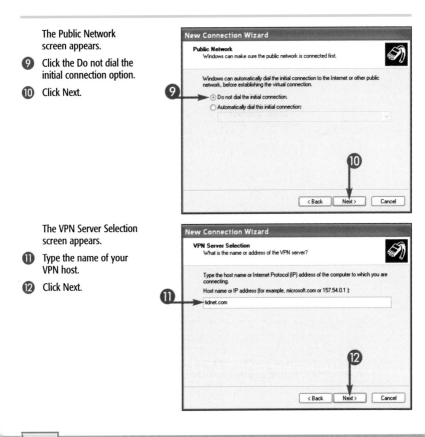

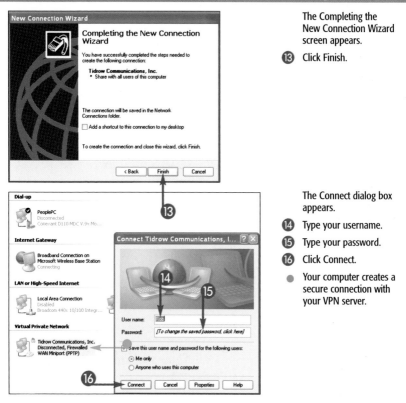

The Completing the New Connection Wizard screen appears.

⑬ Click Finish.

The Connect dialog box appears.

⑭ Type your username.

⑮ Type your password.

⑯ Click Connect.

● Your computer creates a secure connection with your VPN server.

**TIP**

**Did You Know?**
You can disconnect a VPN connection by right-clicking the VPN connection icon and selecting Disconnect.

# Index

# Index

# Index

# Index

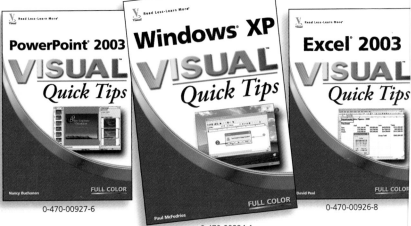